PRENTICE-HALL FOUNDATIONS OF MODERN  SERIES

Sigmund R. Suskind and Philip E. Hartman, Editors

AGRICULTURAL GENETICS
> *James L. Brewbaker*

GENE ACTION, Second Edition
> *Philip E. Hartman and Sigmund R. Suskind*

EXTRACHROMOSOMAL INHERITANCE
> *John L. Jinks*

DEVELOPMENTAL GENETICS*
> *Clement Markert and Heinrich Ursprung*

HUMAN GENETICS, Second Edition
> *Victor A. McKusick*

POPULATION GENETICS AND EVOLUTION
> *Lawrence E. Mettler and Thomas G. Gregg*

THE MECHANICS OF INHERITANCE, Second Edition
> *Franklin W. Stahl*

CYTOGENETICS
> *Carl P. Swanson, Timothy Merz, and William J. Young*

* Published jointly in Prentice-Hall's *Foundations of Developmental Biology Series*.

POPULATION GENETICS AND EVOLUTION

Lawrence E. Mettler
North Carolina State University, Raleigh

Thomas G. Gregg
Miami University, Oxford, Ohio

PRENTICE-HALL, INC.   Englewood Cliffs, New Jersey

FOUNDATIONS OF MODERN GENETICS SERIES

Printed in the United States of America
C-13-685297-1
P-13-685289-0
Library of Congress Catalog Card Number:
69–16809

Current printing:
10 9 8 7 6 5 4 3

PRENTICE-HALL INTERNATIONAL, INC., *London*
PRENTICE-HALL OF AUSTRALIA. PTY. LTD., *Sydney*
PRENTICE-HALL OF CANADA, LTD., *Toronto*
PRENTICE-HALL OF INDIA PRIVATE LTD., *New Delhi*
PRENTICE-HALL OF JAPAN, INC., *Tokyo*

To the memory of Wilson S. Stone (1907-1968)

Foundations of Modern *Genetics*

Genetic research is alive with excitment and revolutionary advances. Important to the development of science and to the evolution of social structure, genetic thought is widening its impact on many areas: immunology, protein chemistry, cellular physiology, developmental biology, medicine, agriculture, and industry.

So many partnerships and such rapidly expanding methodology demand a fresh approach to genetic training—an approach attempted in this series.

The basic principles of genetics are few and simple. We present them with enough description of accessory scientific areas to allow comprehension not only of the principles themselves but also of the types of experiments from which the concepts have evolved. Such an approach compels the reader to ask: What is the *evidence* for this concept? What are its *limitations?* What are its *applications?*

The Prentice-Hall Foundations of Modern Genetics Series presents the evidence upon which *current* genetic thought is based. It is neither a history nor a survey of all genetic knowledge. The short volumes make possible a stimulating, selective treatment of the various aspects of genetics at the intermediate level, and sectional divisions allow free choice of emphasis in differently oriented college genetics courses.

vi

The references cited in each volume and the current research literature are the immediate sequels to this series, but the true sequel remains in the hands of the alert reader. He will find here the seed of more than one enigma, the solution of which he, himself, may help bring into man's comprehension sometime in the future.

SIGMUND R. SUSKIND
PHILIP E. HARTMAN

McCollum-Pratt Institute
The Johns Hopkins University

Preface

A glance at the history of genetics reveals a steady shift to the use of experimental organisms that have been progressively simpler in form and more suited to genetic and biochemical analysis.

Not surprisingly, the first genetic observations were made in conspicuous species such as man himself, and in his domestic plants and animals. These early attempts at understanding inheritance were to a large extent frustrated, however, by the complexity of the organisms, which made it difficult to distinguish between "single factor" inheritance (mendelian inheritance) and "multiple factor" inheritance (quantitative inheritance). Indeed, it is to Mendel's great credit that, using the garden pea, he was able to develop an experimental system that overcame these difficulties and led to the discovery of the basic principles of heredity.

Following the rediscovery of the mendelian principles in the year 1900 the genetic study of such organisms as chickens, rabbits, mice, sweet peas, and evening primroses was largely superseded by the introduction of *Drosophila*, the fruit fly, as an experimental organism in 1910. In the early 1940's the introduction of *Neurospora*, a red mold, was responsible for the next series of major advances. And then in the late 1940's bacteria and subsequently bacterial viruses (bacteriophages), the ultimate in genetic simplicity, were pressed into service. As a result our present knowledge of genetic mechanisms as they operate in the individual, though by no means complete, has become highly sophisticated.

At the same time it has been widely recognized that there are higher levels of organization that are of crucial importance in nature. These are groupings of interacting individuals of the same species, called populations, and groupings of interacting populations of different species, called biotic communities. Consequently, if we are to understand the complexities of nature above the level of the individual and if we are to understand evolution, we must begin by coping with the problems of the population.

The fundamental relationship between gene frequencies and genotype frequencies in populations (Chapter 3) was recognized by the British mathematician H. G. Hardy in 1908. He pointed out this relationship rather apologetically, since it was so simple, in a letter to the editor of *Science*. The German physician W. Weinberg had also struck upon this relationship in the same year. So as our knowledge of the genetics of the individual advanced, population genetics began to develop apace. S. S. Chetverikov, Sewall Wright, R. A. Fisher, and J. B. S. Haldane were without doubt the major figures in the early development of population genetics (roughly between 1920 and 1940), especially of the mathematical and theoretical aspects. Hardly a paper is published on this subject today that does not in some way refer to their work.

Much of the more recent work has emphasized the experimental side of population genetics, a sampling of which is provided in this book along with some discussion of the theoretical aspects. Population studies, especially experimental ones, often have had to await developments in the basic understanding of genetics in the individual. In this respect recently acquired knowledge of genetic mechanisms has opened up new doors, and population genetics appears to be on the threshold of major new advances.

In the preparation of this book we are indebted to many people who have helped or given encouragement in numerous ways. We would especially like to thank Kenichi Kojima and Richard Richardson of the University of Texas, L. C. Saylor, S. G. Stephens, and Terumi Mukai of North Carolina State University, and Harry Weller of Miami University for their many valuable comments and for careful reading of the manuscript.

L.E.M.
T.G.G.

Contents

The Theory of Evolution

Evolution, the process that has produced the fantastically diverse array of organisms alive today, is one of the most basic concepts in biology. From the primitive precellular aggregates of organic molecules that arose in the primeval seas over three billion years ago, to the exceedingly complex and highly integrated multicellular organisms of the present, evolution, and more specifically natural selection, has been the molding force. Therefore, it is not surprising that this concept should pervade every field of biology as the major unifying idea in our understanding of life and organic functions.

In simplest terms, evolution is nothing more than what Charles Darwin called "descent with modification." The modifications result from the differential success in reproduction by individuals possessing different heritable characteristics. Or, as Th. Dobzhansky has defined it, "Evolution is a change in the genetic composition of a population." However, when one considers the lines of evidence for evolution, which are many and diverse, the details of the various mechanisms which give rise to the modifications in different types of organisms, and the various forces and interactions that are involved, the subject becomes vast and complex. Actually there are many facets of it that we do not yet fully comprehend. Indeed, as Darwin himself pointed out in *The Origin of Species*, "Anyone whose disposition leads him to attach more weight to unexplained difficulties than to the explanation of a certain number of facts will certainly reject the theory." This observation still applies today, although there are more explained facts and fewer major unexplained difficulties than in Darwin's time.

The lines of evidence for descent with modification are circumstantial, but when taken together they are overwhelmingly convincing. Today there is general agreement among scientists and, indeed, most people that evolution has occurred, is still occurring, and that it fully accounts for the diversity of existing organisms. In this book we shall not be concerned with reviewing the evidence for evolution, which is more appropriately covered in a more general treatment, nor with presenting the chronological history of descent of living forms. Our discussion will be limited to evolution defined as a process and the forces associated with it; especially those dealing with genetics and genetic mechanisms.

Darwin's monumental work of 1859, *The Origin of Species*, completely changed western man's concept of the natural world. It was the culmination of a long period of groping by naturalists and systematists for a causal explanation of the many similarities between otherwise different kinds of organisms. Darwin's explanation, which appears trite today, was simply that similar species are closely related by descent. In general, the more closely related two groups are the more similarities there are between them, and vice versa. Thus, the central theme of evolution is that all existing organisms are descended from one type or a few types of simple primitive organisms that first appeared several billion years ago. Some authors have even extended this concept backwards in time to include molecular aggregates which had not yet achieved a cellular level of organizational complexity, but which reproduced themselves (and their mutant forms) reliably enough for natural selection to operate. Such an aggregate could then have given rise to new lines of descent. As a consequence, except for instances of convergent evolution, at any level in a phylogenetic tree the similarities shared by various groups have persisted from a common origin in a common ancestral form, whereas the differences have arisen during their subsequent isolation in the history of each group.

Although Darwin was by no means the first to put forth a theory of evolution, he was the first to present one that was comprehensible and convincing.[1] Several things contributed to his success. His careful objective collection and compilation of evidence indicating that species have changed and do change and are not "immutable"—in other words, that evolution does occur—were so extensive as to be virtually irrefutable. Of equal importance was the fact that he was able to

[1] Actually, Alfred Russell Wallace, a naturalist who carried on extensive investigations in the East Indies, formulated a theory of evolution essentially identical to Darwin's at the time Darwin was writing *The Origin of Species*. After a joint presentation of the theory to the Royal Society of London in papers by Wallace and Darwin in 1857, Darwin went on to complete *The Origin of Species* on the merits of which he has been accorded major recognition for the theory's formulation.

present a very logical and convincing explanation for the occurrence of these changes, i.e., natural selection. Again, Darwin was not the first to introduce the concept of natural selection, but he was the first to fully comprehend the relationship between natural selection and heritable changes in populations. Therein lay his triumph; not only could he demonstrate that changes had occurred, but he could explain how they had come about.

We can summarize the basic tenets of the Darwinian theory of evolution as follows (quotations are from *The Origin of Species*):

1. *The number of individuals in any population tends to increase geometrically when conditions permit the survival of all progeny.* "There is no exception to the rule that every organic being naturally increases at so high a rate that, if not destroyed, the earth would soon be covered by the progeny of a single pair."
2. *The potential for rapid increase is seldom realized.* "In the case of every species, many different checks, acting at different periods of life, and during different seasons or years, probably come into play; some one check or some few generally being the most potent; but all will concur in determining the average number or even the existence of the species."
3. *Darwin deduced from these facts that a competition or "struggle" for survival occurs in which many individuals are eliminated.* "Hence, as more individuals are produced than can possibly survive, there must in every case be a struggle for existence. . . ."
4. *Variation, in the form of individual differences, exists in every species or population.* "No one supposes that all the individuals of the same species are cast in the same actual mould. These individual differences are of the highest importance for us, for they are often inherited, as must be familiar to everyone. . . ." Although the mechanism of heredity was confused and essentially unknown to Darwin, the *fact* of heredity was appreciated and served as the outstanding feature for explaining the theory of evolution. "Any variation that is not inherited is unimportant to us."
5. *From the observed differences between individuals as well as those between varieties, Darwin deduced that the elimination process was selective.* "Can it then be thought improbable, seeing that variations useful to man for plant and animal breeding have undoubtedly occurred, that other variations useful to each being in the great complex battle for life, should occur, can we doubt (remembering that many more individuals are born than can possibly survive) that individuals having any advantage, however slight, over others, would have the best chance of surviving and procreating their kind?" The surviving ones are considered to be more fit. But *fitness* is not defined in the limited sense of the organism's relative ability to "struggle" with competitors for food, space, and mates, nor simply by its chances of escaping predation and disease. Darwin recognized that fitness is best defined as the relative capacity for

leaving offspring; ". . . I use this term in a large and metaphorical sense including dependence of one being on another, and including (which is more important) not only the life of the individual, but success in leaving progeny."

6. *Evolution is a gradual change in the hereditary makeup of the species.* "Certainly no clear line of demarcation has as yet been drawn between species and sub-species—that is, the forms which in the opinion of some naturalists come very near to, but do not quite arrive at, the rank of species: or, again, between lesser varieties and individual differences. These differences blend into each other by an insensible series; and a series impresses the mind with the idea of an actual passage.

"Hence I look at individual differences, though of small interest to the systematist, as of the highest importance for us, as being the first steps towards such slight varieties as are barely thought worth recording in works on natural history. And I look at varieties which are in any degree more distinct and permanent, as steps towards more strongly-marked and permanent varieties; and at the latter, as leading to sub-species, and then to species. The passage from one stage of difference to another may, in many cases, be the simple result of the nature of the organism and of the different physical conditions to which it has long been exposed; but with respect to the more important and adaptive characters, the passage from one stage of difference to another, may be safely attributed to the cumulative action of natural selection. . . . A well-marked variety may therefore be called an incipient species."

In this context Darwin pointed out that the existence of numerous examples of "incipient" species in nature is strong evidence for the occurrence of evolution (Fig. 1.1).

Only the essential points of Darwinism have been presented in this brief outline. Natural selection and evolution, or descent with modification and *The Origin of Species,* were corollaries drawn by Darwin from the observed facts of inheritance, variation, high fecundity, and selective mortality.

Our modern ideas about evolution include several features that were not a part of Darwin's theory, but which have been called variously "the modern synthesis," "the neo-Darwinism synthesis," or just "neo-Darwinism." In this synthesis, natural selection is still the major molding force, but our knowledge of the particulate nature of genes enables us to understand more fully the origin of variation by mutation, the preservation of concealed variation in diploid organisms, and the shuffling of the genes by genetic recombination so that new combinations are constantly available for natural selection to act upon. These additions have not altered the central framework of the original theory. Evolution still is seen as a two-part process: the origin of variation and the modification of the variation by natural selection.

We should also point out that the changes brought about by natural

Fig. 1.1. "It is interesting to contemplate an entangled bank, clothed with many plants of many kinds, with birds singing on the bushes, with various insects flitting about, and with worms crawling through the damp earth, and to reflect that these elaborately constructed forms, so different from each other, and dependent on each other in so complex a manner, have all been produced by laws acting around us. These laws, taken in the largest sense, being Growth and Reproduction; Inheritance which is almost implied by reproduction; Variability from the indirect and direct action of the external conditions of life and from use and disuse; a Ratio of Increase so high as to lead to a Struggle for Life, and as a consequence to Natural Selection, entailing Divergence of Character and Extinction of less improved forms. Thus from the war of nature, from famine and death, the most exalted object which we are capable of conceiving, namely, the production of the higher animals, directly follows. There is grandeur in this view of life, with its several powers, having been originally breathed into a few forms or into one; and that, whilst this planet has gone cycling on according to the fixed law of gravity, from so simple a beginning endless forms most beautiful and most wonderful have been, and are being, evolved." (Charles Darwin, *The Origin of Species.*) Photograph courtesy of F. M. Johnson.

selection may have different evolutionary consequences, depending on the circumstances. The first occurs when the environmental factors determining natural selection are fairly uniform throughout the population or species range. In this situation the entire species will tend to become better and better adapted to its environment and, as the environment changes, the entire species will change with it. Given enough time, marked changes can occur in this way. Thus, the genetic composition of a single line of descent is gradually and uniformly altered over succeeding generations. This is called *phyletic evolution*. If we had a time machine so that we could directly compare the characteristics of this species at the beginning and end of a one million year

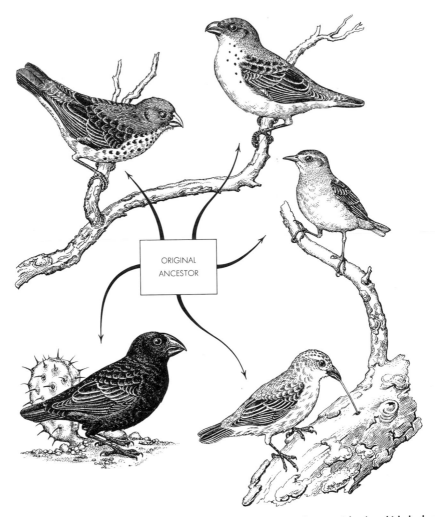

ORIGINAL
ANCESTOR

Fig. 1.2. Some of the different species of finches from the Galapagos Islands, which had a marked effect on Darwin's thinking as he grappled with his theory:

"Of land-birds I obtained twenty-six kinds, all peculiar to the group and found nowhere else, with the exception of one lark-like finch from North America. . . .
The other twenty-five birds consist, firstly, of a hawk, curiously intermediate in structure between a Buzzard and the American group of carrion-feeding Polybori. . . . Secondly, there are two owls. . . .
Thirdly, a wren, three tyrant fly-catchers . . . and a dove—all analogous to, but distinct from, American species.
Fourthly, a swallow. . . . Fifthly, there are three species of mocking-thrush. . . .
The remaining land-birds form a most singular group of finches, related to each other in the structure of their beaks, short tails, form of body, and plumage: there are thirteen species, which Mr. Gould has divided into four sub-groups. All these are peculiar to this archipelago. . . .
Seeing this gradation and diversity of structure in one small, intimately related group of birds, one might really fancy that from an original paucity of birds in this archipelago, one species had been taken and modified for different ends." Figure from Stebbins (1966); after David Lack.

period, we might find that the differences between them were so great that they actually constituted different species. The second consequence occurs when different populations in a species somehow become isolated (see Chapter 8) and are subjected to different environmental conditions. With natural selection acting nonuniformly, different kinds of changes will take place in the various populations. In this way they may become more and more divergent until the single original species has split into two or more new ones. The latter case is the only way in which the number of species can increase and is referred to as *speciation*. Speciation is a momentous occurrence in evolution in that it leads to discontinuities among populations, with each population representing an independent line of descent possessing its own unique potentialities for further phyletic change. Thus, divergence can be a continuous process and need not stop when the species level has been attained, but can and must give rise to higher taxonomic categories as well (Fig. 1.2).

Evolution is a phenomenon that involves genetic changes at the species or population level of organization. These groups are the basic units of evolution, and to understand and explain the forces which bring about changes in them it has been necessary to apply our knowledge of genetics to populations. But we must define these units in meaningful terms before the evolutionary forces operating on them can be discussed. This we shall proceed to do.

References

Darwin, Charles, *The Origin of Species*. (1859; paperback available, New York: Washington Square Press, 1963.)

Eiseley, L., *Darwin's Century*. New York: Doubleday and Co., 1958. (Paperback available: Anchor Books, 1961.)

Grant, Verne, *The Origin of Adaptations*. New York: Columbia University Press, 1963. Part 1, Chapters 1, 2, 3, and 4. Philosophical aspects of science. Causal theories in biology. Abiogenesis and the course of evolution.

Grebstein, S. N., *Monkey Trial*. Boston: Houghton Mifflin, 1960. Houghton Mifflin Research Series, Number 4. This is a well-researched account of the Scopes trial in Tennessee. Many philosophical issues are covered. Very light reading.

Nogar, F. J., *The Wisdom of Evolution*. Garden City, N.Y.: Doubleday and Co., 1963.

Stebbins, G. Ledyard, *Processes of Organic Evolution*. Englewood Cliffs, N.J.: Prentice-Hall, Inc., 1966, p. 148.

Two

Decoding the Diversity of Life

"What is a species?" To this interesting question we shall now attempt to find an answer. It may be no tidy answer, and we shall doubtless find a host of equally intriguing questions.

The scientific art of ordering organisms

Life appeared on the planet Earth more than 3 billion years ago and has evolved since then to the tremendous array of organic forms living today. More than a million forms of animals and more than a quarter million types of plants have become known through the efforts of naturalists and systematists of the 19th and 20th centuries. In addition, a multitude of extinct forms has been unearthed by paleontologists. It has been crudely estimated that the total number of distinct kinds of organisms throughout the history of life has been more than a billion. Possibly as many as 415 million of them are still in existence. Although certain classes of organisms, such as birds and mammals, are well cataloged, numerous types surely remain to be discovered or formally recognized, especially among the insects, the class already including by far the greatest number of described forms.

Most people are familiar with the basic aspects of this diversity and can point out some of the differences between individuals of different groups. Reptiles are easily distinguished from birds; the category designated "cat" is never confused with that called "dog"; lions constitute a group distinct from tigers. We also recognize variation among members of a single group. It is obvious that every individual is unique and that no two persons, and presumably no two organisms, are ever exactly alike; every mother knows her child, and every child knows his pet. Although differences exist among groups and among individuals within groups, similarities, based upon propinquity of descent, are also evident to most of us. The awareness of similarities between individuals is expressed by such time-honored statements as "like begets like," "each after his kind," and "birds of a feather."

The existence of similarities among individuals and differences between groups has enabled man to systematize the organic world into orderly patterns. Categories of living organisms have been recognized and named in the vernacular throughout the world. In addition, formal names have been applied to groups which have been scientifically ordered. The science of systematics, or taxonomy, is the study of the myriad of organisms: the kinds, their distinctive features, and classifications, which are determined by the study of any and all relationships among them.

In the early days of this science, classification was simply the act of discovering the group, determining its diagnostic characteristics, and naming it. Each kind of organism was believed to be independently created, conforming to a single morphological pattern over time and space, and easily separable from other forms. Classification was a matter of pigeonholing morphologically distinguishable "true" entities, eventually called species, into some "natural" system. This was conceived of as a relatively simple chore, since species were assumed to be composed of individuals which conformed to a single type, as if cast in the same mold. Individuals that varied from the type were considered exceptions and were excused simply as mistakes, or "sports," in development. The shared attributes among different species, or affinities between species, were vaguely explained as resulting from the whim of nature or of the supernatural during the independent creation of each type.

Aside from simply naming groups, discovering the natural system, or "plan of creation," was the primary aim of early systematists. Darwin gave what is now considered a most satisfactory answer to the problem. Thanks to him, the evolutionary history of populations eventually became the rational basis of classification, and the phylogenetic interpretation of relationships between different organisms served to elucidate the natural system. He suggested that the hierarchic classification system of taxonomy, i.e., grouping related populations into one species, combining related species into the same genus, assigning similar genera to the same family, families to orders, orders to classes, etc., is a "natural subordination of organic beings in groups under groups" and is not arbitrary like grouping stars into constellations. The hierarchic classification reflects a reality of nature, a natural system, which is based upon different degrees of phylogenetic divergence during evolutionary descent. The hypothesis suggests that present-day species arose from common ancestral forms at different times in evolutionary history, that there is a tendency for a divergence of the characteristics of separate forms through gradual change, and that the degree of divergence (the extent of difference between groups) reflects the remoteness from the common ancestral form. Actually applying the concept of phylogeny to classification, however, is most difficult.

The difficulty of establishing a natural system rests with interpreting the available evidence in such a way that groups are categorized phylogenetically. Kinship in evolutionary descent is the fundamental criterion. This kinship is usually inferred from similarities of the morphological and anatomical characteristics of the groups studied. The data seldom include direct evidence, i.e., known reproductive interconnections in descent. In fact, a large number of the known kinds of organisms have been described and named after examining a few preserved specimens in museum collections. In relatively few instances has the population structure, breeding habits, ecological requirements, cytological and physiological characters, or distribution of a group been well documented, and seldom has a group been sampled adequately to reveal its total variation. But regardless of the number of observed relationships, phylogeny is a postulation. Even the phylogenetic sequence in the long graded series of fossils provided by paleontology was made possible only through inferring their reproductive interconnections, which of course were not observed.

The natural classification system of modern taxonomy is based upon phylogeny, which in turn rests upon the evolutionary principle of divergence in descent. Categories in any such classification are therefore defined in evolutionary terms, but the evidence used to establish categories is almost entirely that of similarities between forms; actual descent is usually inferred. Simpson (1961, p. 69) helped clarify this confusing distinction with an example of monozygotic ("identical") twins: "We *define* such twins as two individuals developed from one zygote. No one has ever seen this occur in humans, but we recognize when the definition is met by *evidence* of similarities sufficient to sustain inference. The individuals in question are not twins because they are similar but, quite the contrary, are similar because they are twins." Suggesting that a biological classification built upon evolutionary principles is for that reason "natural" does not mean, however, that only one system is possible, nor that any one category objectively delimits a real unit of nature. One systematist might call a group a genus, while another might prefer to list it as a family or as two genera. Leading systematists of today, however, do assert that the basic category of classification, the species, actually does reflect distinct entities, much as individuals, cells, or chromosomes are distinct organizational entities.

The species category

The systematist is confronted with a vast array of biological types that could be ordered into many kinds of classification systems. Nevertheless, if for no other reason than to standardize nomenclature, it is necessary that they conform to one official system. For the past two centuries Linnaeus's hierarchy (expressed in *Systema naturae*, 1758 edition) has been accepted in its general form by the zoologist and in its essence by the botanist. The basic unit of this system is the species. By tradition, every organism is assigned both to a species category and to higher, more general supraspecies categories (genus, family, order, class, phylum, etc.). Every individual is named by the binomial method according to the genus and species to which it belongs. Systematists have yielded to this tradition and have been constantly confronted with the problems of identifying species and of naming them in a manner consistent with some definition of the species category. This has led to the so-called "species problem." How can the species category be objectively defined? In other words, what characteristics are to be considered diagnostic and how much differentiation must exist between populations before they are to be recognized as separate species? The problem has become understood (Mayr, 1963; Simpson, 1961), but an objective definition which includes criteria for universal application is still desired. It is now realized that criteria for *recognition* (practical level) can never be established as long as the species definition is based on evolutionary concepts and involves dynamic systems (see page 17). This should not deter us from *defining* (theoretical level) the species category by some reasonable and objective criterion. In the following sections we shall attempt to find such a criterion.

The evolutionary species concept

Simpson (1961) has defined an evolutionary species as "a lineage (an ancestral-descendent sequence of populations) evolving separately from others and with its own evolutionary role and tendencies." Immediately we see that this is an abstractive rather than a practical definition: its application is difficult if not impossible because of the lack of rules on which to base an undisputed description of "evolutionary role and tendencies." But it is a most inclusive definition—a theoretical standard to which operational definitions may be compared.

The species as here conceived is a discrete, reproductively defined system of populations with common evolutionary history. Such histories are diagrammatically represented by "trees," which are a pictorial interpretation of an actual phylogeny of descent (Fig. 2.1). In such trees, the terminal points of all lineages that reach the present

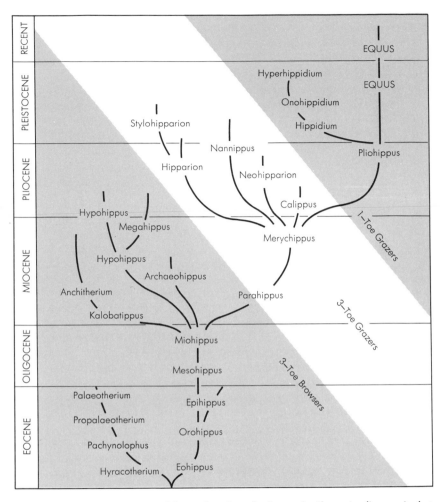

Fig. 2.1. The phylogenetic tree of horses based on fossil records. The major lineage is that from *Eohippus* to modern *Equus*. Many side branches are extinct. Evolution of the major lineage occurred mostly in North America. *Hyracotherium* (presumably the same as *Eohippus*) and its descendants occurred in the Old World, as did *Anchitherium*, *Hipparion*, and *Stylohipparion*; the *Hippidium* lineage from *Pliohippus* occurred in South America. Only those forms of *Equus* that spread during the Pleistocene age to Eurasia and Africa gave rise to present-day horses, asses, and zebras. Phylogenetic interpretations after Matthew, Loomis, and Simpson.

on the time scale are living species, and terminal points in the past represent extinct populations. The connections of the lineages through time are generally inferred from interpretations of the relationships of *existing* groups, although fossil records help to increase the probability of a correct description of the phylogeny.

If a complete paleontological record were available, a single evolutionary lineage (called a gens) could be traced back through phylogenic history as a series of ancestral-descendent populations. But this does not mean, for example, that the sequence tracing backward from *Equus* down through *Pliohippus, Merychippus, Miohippus, Mesohippus*, and *Eohippus* constitutes a single species. According to the theory of phyletic evolution, enough changes might occur in the course of time that there is a transformation of one species into another during descent within the same lineage. Many species must have been involved in the gens from *Eohippus* to *Equus*. Given the principle of sequential species formation by phyletic evolution, we divide any such lineage into successional species, or paleospecies (Cain, 1954). And here we encounter one aspect of the species problem: keeping in mind the evolutionary concept of species, just how much difference between ancestral-descendent populations (i.e., between two fossils) is necessary before we label them two species? This certainly would be a subjective decision if a complete record were available and transformation were of a gradual nature. In practice, however, the recognition of paleospecies is facilitated by gaps in the fossil record, which show clear-cut species on either side, and by places where lineages separate.

According to the evolutionary concept, a species is considered as a sequence of ancestral-descendent populations constituting a segment of a single lineage. Other concepts of the species which emphasize one or more of the relationships in *existing* groups of organisms have been set forth without particular reference to evolutionary histories. Mayr (1963) lists three such concepts: (1) the typological concept; (2) the nondimensional concept; and (3) the multidimensional concept.

The typological concept (morphological species)

The basic ideas about species have had their roots in two opposing hypotheses, constancy versus transmutation of organisms, which have alternated in importance in the minds of men throughout history. As pointed out above, evolution presently is considered to be the process leading to the origin and continuity of species. This latest stage in the development of systematics was ushered in by Lamarck, but it was not firmly established until Darwin's publication of *The Origin of Species*. During the hundred-year period before this publication, Linnaeus's influence was supreme.

In the Linnaean hierarchy, the species was the lowest rank above the individual. The individuals of a species were considered to be relatively uniform in anatomical features and clearly separable from other kinds of life. Essential to this idea was the belief that each species was separately created and that each has maintained constant morphological characteristics, varying only within prescribed boundaries established by the Creator.

Since it was believed that the species consisted of individuals built from a common "plan," a single typical specimen could be used to analyze the essential features of the total group. These ideas, emphasizing static species, are the basis of the typological concept. *The principal criterion for describing species according to this concept rests primarily with various degrees of morphological differences.* Species so described are referred to as morphological species, or morphospecies (Cain, 1954).

Two factors were responsible for the acceptance of the static morphospecies by the early systematists: first, universal support was given to the literal interpretation of the story of divine creation expressed in Genesis and second, species were customarily described from the study of a few individuals collected from a single locality, little attention being given to the local or geographic variation among the individuals constituting the group.

The nondimensional concept

The nondimensional concept might best describe the idea of species held by local naturalists, who observe populations of different kinds of organisms coexisting in the same place at the same time. For example, an amateur ornithologist might observe several kinds of birds (such as cardinals, bluejays, juncos, starlings, white-throated sparrows, and song sparrows) at his feeding stations all within a relatively short period of time. Each kind is distinct from the next, and most if not all individuals are easily identified according to kind. Each distinct population is a particular species.

In order to understand this concept fully, it will be helpful to add to our terminology. Conspecific individuals (members of a species) of cross-fertilizing forms are associated in populations. At any given time any species has a limited range in space, a range composed of the total areas occupied by all its populations. The extent of any range and the size and spatial arrangement of the breeding colonies within depend on: (1) the mating system; (2) the vagility, or dispersal facility of the species, which may be active (individual mobility) or passive (conveyance by some other agent—wind, water, or other organism); (3) the distribution of the ecological conditions that make up the requirements for life of the species (climatic and biotic barriers to range expansion); and (4) the nature of the geographic barriers to dispersal (such as oceans, mountains, rivers, lakes, glaciers, and deserts). Considered over time, a range is dynamic; its edge changes in a fashion similar to that of the edge of a body of water. Under certain conditions the species may flood an extensive area, under others it may be limited to a small, precisely defined territory. A highly dispersive species, generally adapted to a spectrum of environments, is usually uniformly spread over its range. However, the populations on the margin of the range often exist as semi-isolated "pools," which unite with the rest or separate as the range is altered. Less dispersive forms, which are narrowly adapted to specific conditions, may consist of small breeding colonies throughout the total range.

The ranges of different species may or may not overlap. Individuals of all species that coexist in the same geographic territory are referred to as sympatric. Individuals, populations, or species occupying different geographic regions are said to be allopatric. Allopatric populations may be contiguous (in juxtaposition), or they may be disjunct (separated by a wide intervening distance or by geographic barriers). Most contiguous populations are specifically adapted to adjacent habitats—they are geographic neighbors. They are considered geographically allopatric, *but* if individuals or reproductive cells come into contact along their common margin, so that interbreeding is possible, the populations are regarded as sympatric in a genetic sense. We shall follow Grant's (1963) suggestion, and call contiguous populations *adjacently* (or neighboringly) *sympatric*, and those populations occupying different niches within the same habitat *biotically sympatric*. If neighboring or biotically sympatric populations do not exchange genes by intercrossing, they are considered separate species, whereas gene exchange, evidenced by the existence of intermediate forms, indicates that they are populations of one species.

Each species present sympatrically in a restricted local area is represented by a single population (see Chapter 3) which does not reveal geographic variation and racial divisions. Local populations of different species are by definition independent interbreeding units, and most often they can be clearly defined and separated from each other by the presence of morphological discontinuities, or "species gaps." For example, Hartman's zebra in the mountain region in South-West Africa, from the Kaoko Veld of Damaraland to the Orange River, is easily identified by certain gross morphological features. In the Etosha Pan area in northern Damaraland, herds of these zebras have been observed overlapping the western part of the range of the equally well-defined form called bontequagga, or Chapman's zebra (Fig. 2.2). These two species are morphologically separable by size, color, striping pattern, habits, and cries. An obvious discontinuity exists between them. The distinctiveness of their populations is emphasized by the associations of the individuals into species-specific herds in the zone of overlap. The same kind of "species gap" exists between Grevy's zebra and Grant's zebra. Their ranges merge on the plains between Lake Rudolf and Mount Kenya in Kenya. Grevy's zebra is much the larger of the two and possesses narrower and more numerous stripes. Likewise, two distinctive forms, the nearly extinct Cape Mountain zebra and the Quagga (extinct since 1870), were occasionally seen coexisting in a small area south of the Orange River in Cape Colony a century ago. All six zebras mentioned above were named before the turn of the century, without a knowledge of their ranges, as separate morphospecies. Another extinct form, Burchell's zebra, found north of the Orange River, apparently existed allopatrically with the Quagga and was also given specific rank. In these cases the "species" are characterized by the *noninterbreeding* of two coexisting local populations. This constitutes the nondimensional concept of species.

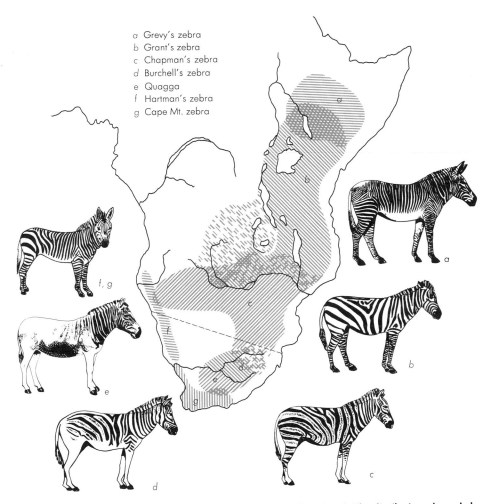

a Grevy's zebra
b Grant's zebra
c Chapman's zebra
d Burchell's zebra
e Quagga
f Hartman's zebra
g Cape Mt. zebra

Fig. 2.2. The distribution and striping patterns in zebras (a—g). The distributions shown below the dashed line are based on old records (prior to 1900) of zebras now extinct, or nearly so. A cline in striping characteristics is exhibited between Grant's zebra, Chapman's zebra, and Burchell's zebra (and possibly including Quagga). These forms constitute races of one multidimensional species, *Equus burchelli*, which is distinct from Grevy's zebra (*Equus grevyi*) in the north and the mountain zebras (*Equus zebra*) in the south.

The multidimensional concept

After naturalists began to travel more extensively, they soon discovered that between many local populations that appeared quite distinct in their specific localities were morphologically intergrading groups. A single population could no longer be considered necessarily a separate species. It was then realized that "the species" is a group phenomenon (a multiple-population system distributed over a geographic range). Interpopulation variation could not be ignored when describing such a system.

Let us reexamine the zebra example. The seven zebras named as morphospecies did not form phenotypic intergrades in the three regions where and when their populations were sympatric. Each bred true (no hybridization). However, with descriptions of additional populations located in the more remote regions, it has become known that populations of Grant's zebra, Chapman's zebra, and Burchell's zebra intergrade morphologically and replace each other geographically. They form one widely dispersed population network with no apparent morphological "species gap." A weak morphological gradient in the character of striping can be demonstrated among the populations from the north to the south. Grant's zebra is boldly striped with broad distinct rays extending to the hoofs. Starting with the populations south of the Zambezi River, the pattern is less distinct, and some bands form faint interrays (Chapman's zebra). There is also a reduction of pigmented bands on the legs. The extinct Burchell's zebra farther south had almost pure white limbs and undersurface. Geographic barriers, such as rivers (Zambezi, Orange, and Vaal) and desert regions (Bechuanaland), have prevented a completely free genetic interchange among these populations. For this reason, the character gradient is not completely smooth but resembles a variation pattern characteristic of geographically isolated populations. Such character gradients, or clines, are morphological evidence that the contiguous populations are genetically related, forming a single higher-ordered population system —a single species. This evidence has led to the grouping of our three locally defined zebra "species" into one multipopulation unit called *Equus burchelli*.

Quagga is an enigma. The extreme reduction of striping and the light undersurface would logically place these creatures at the terminus of the *E. burchelli* color cline. Indeed, some taxonomists have placed them in this species. But others have felt that Quagga was too distinct to be grouped with the more boldly striped forms. In addition, there are old reports that Quagga and Burchell's zebra roamed together without interbreeding north of the Orange River. A species gap would be indicated if this were true. However, other reports are contradictory, and Quagga may well have been allopatric, in which case its species assignment would become a subjective decision.

Geographic variation in *E. burchelli* led to the naming of several local nondimensional morphospecies. This is also true of *E. zebra* (Hartman's zebra and the Cape Mountain zebra). Species like these, which exhibit geographic variation and are composed of identifiable subordinate groups (or races), are called polytypic. The species *E. grevyi* has a more restricted range and does not vary geographically in taxonomically important features. It is monotypic. Since the discovery of abundant examples of polytypic species in all major groups of plants and animals, emphasis has shifted from characterizing individuals and local populations to describing population systems. Let us now attempt to define the species category in a way consistent with the multidimensional concept.

Defining species under the multidimensional concept

The early ideas about species were formulated out of the practical need for naming and identifying different kinds of organisms. Definition of species necessarily had to be universally applicable for all forms encountered. The morphological definition served the purpose. Today, practicing systematists are still concerned with the identification of individuals and the naming of species, but the conceptual background has changed. Although ascertaining morphological differences among individuals is still necessary—indeed, it is the outstanding aspect of taxonomic procedure—a species is now conceived of as an assemblage of variable populations dynamically evolving as a collective unit reproductively isolated from the rest of the biotic world. Now that it is realized that a single reproductive community can be composed of morphologically distinct subunits, the primary criterion for defining species has become the reproductive relationships among populations rather than the morphological differences between individuals.

The reproductive criterion is the basis of the so-called modern biological definition. This definition of species includes the following four essential points:

1. Biological species occur only among sexually reproducing forms.
2. A species consists of one or more populations of interrelated individuals (in accordance with the multidimensional concept).
3. The interrelation of the members of any population of a species results from their dependence on one another for reproductive purposes. (Disjunct populations of the same species are considered to be potentially capable of gene exchange.)
4. The ultimate criterion of a biological species is that its populations are reproductively isolated from all others. (It is this aspect of species that accounts for the species gap under the nondimensional concept, and the "separate evolutionary role and tendency" under the evolutionary concept.) In short, biological species are "groups of actually or potentially interbreeding natural populations which are reproductively isolated from other such groups" (Mayr, 1940).

It should be clear that sympatric, morphologically distinct populations of a species cannot exist for long because interbreeding and gene exchange would result in an integration of their separate characteristics. It follows that distinct sympatric populations represent separate species and that they are distinctive because they do not successfully hybridize. Man and certain of his domestic breeds and crops might be considered exceptions. Because of imposed checks to free crossing, several breeds (races) can be maintained together without the loss of their separate characteristics. For example, a dog breeder may keep many kinds of pedigreed stock in his kennel but, because they are potentially capable of interbreeding, he must at least keep them in different pens during the breeding season in order to prevent mongrelization. Human races too can exist sympatrically and still be genetically isolated, because cultural differences form barriers to free intermarriage. But social attitudes cannot completely control the mating urge, and some intermatings always occur whenever human races come in contact with one other. Because of actual or potential interracial reproduction, then, all human races belong to but one species according to the biological definition.

The significant feature of the biological species is reproductive isolation. For any of several reasons forms may fail to mate with one another or to produce viable and fertile offspring which can contribute genetically to succeeding generations. The most obvious reason is simply that the individuals are separated over too great a distance. Geographic separation is purely extrinsic and need not mean that the disjunct forms are not potentially capable of intercrossing. In contrast, populations may live contiguously or biotically sympatric and still not effectively hybridize. Gene exchange between the populations is prevented by one or more intrinsic barriers called isolating mechanisms (see Chapter 8). Isolating mechanisms include such factors as the failure of two forms to mate or cross-pollinate, ineffective fertilization, the inviability of hybrid progeny, or hybrid sterility. According to the biological definition of species, populations belonging to the same species interbreed and produce viable fertile offspring, or if the populations are disjunct, we infer that they are *potentially* capable of effectively interbreeding; populations of different species, on the other hand, are reproductively isolated by one or more isolating mechanisms.

Limitations of the biological concept of species

The biological species definition applies to the evolutionary concept at any particular time in descent. Phylogenetic lineages remain distinct so long as their populations do not genetically commingle. Let us again emphasize that these concepts apply objectively to *defining* the species category but the *evidence* that the definition is met in practice is usually indirect, and is more than likely quite subjective. Just as the reproductive interconnections between successional species (intergrading fossils) are inferred, the reproductive relationship of contemporaneous forms also is often judged by indirect evidence. Three major types of observations are used to ascertain if two forms (populations) should be ranked as different species:

1. The forms are morphologically so distinct that they could hardly be considered one species. It is inferred that the degree of reproductive isolation is correlated with the amount of distinction. The decision of the systematist in such cases is obviously subjective.
2. Representative members of the two forms are placed together experimentally but fail to mate, or if hybrids are formed, they are weak or sterile.
3. The forms exist adjacently or biotically sympatric but maintain their distinctiveness; there is no evidence of crossbreeding or the formation of hybrids (a species gap exists).

Only the third qualification is anywhere near definitive. Even with this rigid yardstick, however, difficulties often arise in naming species. Three major kinds of difficulties are found.

First, *the evidence is meager, inconclusive, or contradictory.* For example, we have seen previously that Quagga may be considered an extreme form of *E. burchelli* or a separate species. Here test 1 is inconclusive, test 2 is lacking, and test 3 has been questioned.

Some of the most difficult cases to decipher are those in which there is a discrepancy between tests 1 and 3. The degree of reproductive isolation between populations is not always reflected by the amount of morphological difference between them. For example, hybrids between the two closely related species *Drosophila pseudoobscura* and *D. persimilis* have not been found among the thousands of individuals that have been examined from samples taken from the area where their populations are sympatric (Fig. 2.3). Exchange of genetic material between the two forms apparently does not occur (test 3). It is obvious that they have embarked on separate evolutionary pathways and, as defined earlier, that they must be separate species. Yet they are so similar morphologically (test 1) that for many years they were thought to be identical and were sorted only by "mating-type" tests conducted in laboratory cultures (test 2). Species such as these, which are exceedingly similar morphologically, are referred to as sibling species.

It is now known that some subtle features which can be used by specialists to distinguish the two species actually do exist. These include differences in male genitalia, sex combs (stout bristles on the front legs), wing shapes, and chromosomes. In addition, the species differ in their ecological requirements and ranges.

Many examples of sibling species have been reported for all the major animal and plant groups. They are very common among the insects, especially in the better-known genera, such as *Drosophila*. Because of their subtle differences and because the exacting methods needed to detect them can be applied to only a limited number of popu-

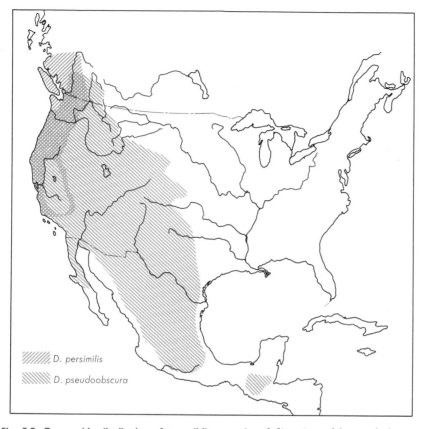

Fig. 2.3. Geographic distribution of two sibling species of flies, *Drosophila pseudoobscura* and *D. persimilis*. The two species are sympatric over a large area from British Columbia to the mountain regions of California. From T. Dobzhansky, *Evolution, Genetics, and Man* (New York: John Wiley & Sons, 1955), page 170.

lations, it is reasonable to assume that many more exist than have been discovered. Polyploidy (multiplication of chromosome sets), a common occurrence with species formation in plants, has resulted in cryptic types in many families of the higher plants. In certain localities two or more diploid species may be found together with polyploid species derived from them through hybridization, all of which are morphologically indistinguishable. As a group they comprise a polyploid complex of sibling yet reproductively isolated species. A morphologically defined species is lost in such a complex.

It must be remembered that morphological differences between groups are the observable expressions of underlying genetic differences operating under environmental influences during development. It is the reproductively independent evolving unit that the taxonomist tries to delimit when defining populations morphologically and giving them the exalted rank of species. Often the amount and kind of genetic distinction necessary to isolate a group of organisms reproductively is also instrumental in changing visible body structures. An experienced taxonomist can often guess correctly the genetic distinction by interpreting the correlated morphological variation patterns. Many morphospecies so named are truly biological species. However, the correlation does not always fit, for there is every degree of anatomical distinctiveness among closely related species. Some show differences so great that one species can be distinguished from another at a distance (Grant's zebra versus Grevy's zebra). Other species may be separable only by special tests (sibling species).

The most perplexing cases are those of related but differentiated allopatric populations that cannot be tested directly for reproductive isolation (test 3). Consider again the two species *D. pseudoobscura* and *D. persimilis*. It has been shown that they will hybridize in laboratory cultures and will produce vigorous F_1 progeny. However, the male hybrids are sterile. Although the female hybrids are fully fertile, their backcross progeny are poorly viable. Gene exchange between the species is effectivey prevented (test 2). These facts support the more important knowledge that hybrids are not formed in the zone of sympatry (test 3), which is conclusive evidence that they should be called species—despite their sibling nature.

All is not so clear-cut with the closely related desert forms, *D. mojavensis* and *D. arizonensis;* the *mojavensis* form consists of three allopatric population groups: in southern California, on the Baja California peninsula, and in Sonora on the Mexican mainland. The related form, *arizonensis*, has been found primarily in Arizona and Sonora (Fig. 2.4). The two forms are biotically sympatric in Sonora and are reproductively isolated (test 3), which is evidence that they are different species. It becomes a much more subjective decision, however, whether or not to place the disjunct California and Baja California populations in the *mojavensis* species. Laboratory tests reveal that members of the three allopatric *mojavensis* populations intercross readily and produce abundant, viable, and fertile offspring (test 2). This evidence, although indirect, indicates that the disjunct populations belong to the *mojavensis* species. Further tests show that members of the disjunct *mojavensis* populations will hybridize with *arizonensis* in the laboratory and produce a small number of viable and fertile progeny. Moreover, backcross and F_2 progeny survive, and certain types are even adaptively superior to both parental forms. It is a moot question whether effective hybridization would occur between *arizonensis* and the allopatric *mojavensis* populations if they were to become sympatric. Those *mojavensis* populations in Sonora apparently do not. It might be mentioned in passing that the California population of *mojavensis* is morphologically distinct from the other two, which happen to be near sibling in nature to *arizonensis* (test 1). All things considered, this is a case of very closely related species, or borderline forms, possibly in the act of speciating. The trouble here is that taxonomists may name one, two, or more species among disjunct populations. They named two. But to insure the recognition of the ambiguity involved, speciation workers are beginning to accept Mayr's terminology, calling allopatric "borderline" forms semispecies and the group of related semispecies a superspecies. Using such terms does not solve the naming problem, but it does allow that the problem exists.

The second difficulty in naming biological species is that *the groups involved are in a state of evolutionary intermediacy.* The process of evolution is gradual. This means that it is difficult in some instances to draw an exact line separating what we call a race from what is a species. Ecological, morphological, and reproductive distinctions among contemporary populations may be only partially developed. Sibling species, for example, have acquired reproductive isolation but display

Fig. 2.4. The distribution of the closely related flies, *Drosophila mojavensis* and *D. arizonensis*. *Mojavensis* consists of three major disjunct populations; the one in southern California is a morphologically distinct geographic race (semispecies). Although members of the two species will effectively hybridize in the laboratory, they are biotically sympatric and reproductively isolated in Sonora, Mexico.

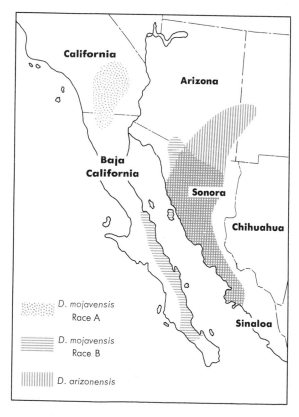

few corresponding morphological differentiations. Other groups may have morphological and ecological characteristics developed to such an extent that they are obviously good morphospecies, but they hybridize freely in the laboratory or experimental garden. This is true of the Eurasian plane tree (*Platanus orientalis*) and the sycamore of the United States (*P. occidentalis*), which have been geographically isolated for millions of years. Still other forms may consist of geographic races with partially evolved reproductive isolation.

The situation becomes even more complicated when the distinctions are not developed to the same extent in all populations of a species. An interesting and well-analyzed case is that of the red-eyed towhee (*Pipilo erythrophthalmus*), which ranges through the United States and south into Mexico and is associated primarily with oaks and bushy undergrowth. In the South-central Plateau its populations are interspersed with those of the collared towhee (*P. ocai*), a more southern species, inhabiting coniferous woods. In this region several populations exist which exhibit mixtures of the characteristics of both species and which intergrade to such an extent that it would be justifiable to think of the two "pure" forms as but one species (Fig. 2.5). However, in at least one locality (Oaxaca, Mexico) the species are known to live sympatrically and do not intermix. It has been suggested that these species have until recently maintained their integrity through reproductive isolation of an ecological nature. With changes in the environment, largely forest clearing by man, novel and intermediate habitats are now available that are suitable for both species, their hybrids, and subsequent descendants. In contrast, the vegetative zones are still intact and separate in Oaxaca.

A host of other examples of hybridization between species have been reported, especially among plants. The fact that some species might hybridize to some extent is not necessarily a contradiction of the biological concept. There is little doubt that contiguous populations that exchange genes freely in their area of contact are conspecific, while sympatric forms that are completely isolated reproductively are different species. But the reproductive isolation criterion cannot always be applied so objectively. After all, reproductive isolation is a population characteristic, which is usually acquired by gradual evolutionary change. Some populations may therefore possess incompletely developed barriers, that is, intermating may occur only occasionally or hybrids may be produced, survive, and be fertile, but not to the same extent as offspring from conspecific matings. Moreover, well-developed isolating mechanisms may break down, allowing some effective intercrossing; what were once considered good species (completely isolated) might then become borderline cases (partially isolated). Hybridization between species most generally consists of but few poorly viable F_1 individuals and some select backcross progeny. Only a small portion of the genes of one species can bridge the "hybrid gap" to the other, and thus the "evolutionary role" of each species remains relatively

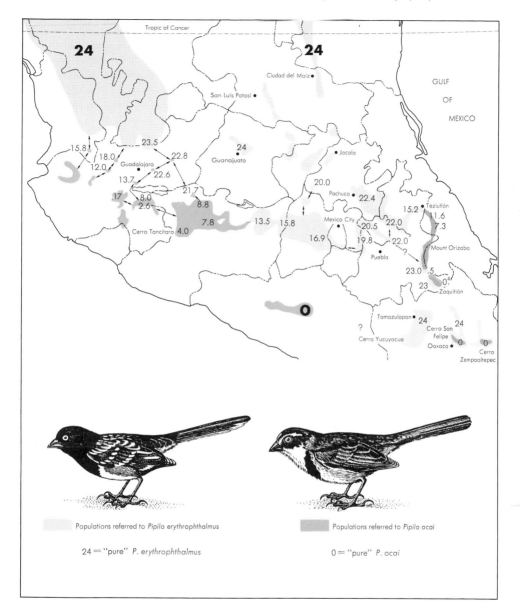

Fig. 2.5. The distribution and integradation of towhee (*Pipilo*) populations in Mexico. The numbers indicate the mean character indices of the various hybrid populations, ranging from "pure" *P. erythrophthalmus* (24) to "pure" *P. ocai* (0). Note the forms remain distinct sympatrically near Oaxaca in the southeast. From the work of C. G. Sibley in *Evolution,* 8 (1954) and in *Univ. Calif. Publ. Zool.,* 50 (1950).

discrete. Such a process of gene infiltration has been termed introgressive hybridization, or introgression. Introgression seems to occur most often between species isolated by ecological factors, as in the case of the towhees. Isolation based on ecology is a fairly weak mechanism for maintaining the distinction of populations and can be destroyed rather quickly by slight environmental disturbances. When introgressing forms tend to maintain their distinction, it may be concluded that they are still different species.

Conspecific populations are characterized by actual or potential interbreeding, while species exhibit reproductive isolation. Contiguous populations showing intermediate amounts of gene exchange by hybridization may be races with partially developed reproductive isolation or species with partly broken-down barriers. In either case the groups cannot be legitimately ranked as races or species. It is convenient to extend the term semispecies to these borderline forms as well as to the allopatric cases.

Semispecies are a taxonomic headache, but they do not represent a frontal attack on the biological concept. On the contrary, they emphasize the dynamic nature of evolving units.

If we could find a truly objective, all-encompassing, and practical criterion for the species, evolutionary theory would surely be in jeopardy. The reproductive criterion has been given special attention because it reflects a basic biological characteristic of sexually reproducing organisms (their genetic independence in evolutionary descent) and thus must be considered nonarbitrary when applied to such units.

The third problem with the biological species definition is *the inapplicability of the criterion to asexual forms*. The two previous limitations to classification and nomenclature were practical ones. Asexual forms constitute a theoretical problem. The "interbreeding-isolation" criterion can apply only to sexually reproducing organisms. Sexual reproduction is defined as any method of replication achieving the important function of combining and recombining genes derived from different familial lines of descent. This is most commonly accomplished by two interdependent processes: (1) gene recombination during meiosis and gamete production; and (2) syngamy, or the fusion of two haploid gametes to form a single diploid zygote. As a result of these processes, genetic information is pooled among contemporaneous individuals and between generations. A higher organizational unit, the population, is thereby established. But by definition no genetic recombination, no genetically defined population structure, and thus no biological species formation is possible among obligate asexual organisms. One of the pressing problems of systematics is the incorporation into the framework of evolutionary theory of the many "parasexual" mechanisms of gene recombination now known in lower forms once considered asexual. Little work has been done to elucidate the population system of such organisms.

In summary, we have come to recognize that the term *species* is used in two ways: (1) *generally,* as the taxonomic designation of the various sorts of organisms that are ecologically clustered and morphologically distinguishable, including asexual organisms; and (2) *specifically,* as reproductive isolates, each composed of populations of actually or potentially interbreeding individuals. The first use of the word is simply a convenient cataloging procedure. Admittedly it results in lumping or splitting evolutionary lineages in many cases, but it is universally applicable and useful for describing grossly the order observed in nature. The second meaning is conceptually sound and serves as a working abstraction from which population phenomena may be generalized, but it is difficult to apply in most instances because of the lack of the necessary facts on which to make an unequivocal decision.

Intraspecific categories (races and subspecies)

Under the multidimensional and evolutionary concepts of species we have seen that a species is composed of populations distributed in space and time. Each local population of a species may be viewed as a colony of interbreeding individuals, which have become adapted to a particular set of conditions in their habitat. Related colonies in different localities in the same general region most likely will occupy somewhat similar niches. Individuals within a local population mate predominantly with one another simply because of proximity, but neighboring breeding units will have a certain amount of dispersal of individuals or reproductive cells between them. Adjacent populations are expected therefore to have most of their genes and characteristics in common. There is less gene exchange between populations of a species that are more dispersed and disjunct. It is also likely that distant populations will become more uniquely adapted to dissimilar environments; they will tend to possess different gene assemblages and different characteristics. Although every population of a species is a unique genetic-physiological response to a local environment, a group of related populations found in the same region will have many shared attributes, but may differ from other groups in other regions in certain characteristics or frequencies of genetic variants. Because of similarities and differences between them, it is sometimes convenient to categorize populations into intraspecific groups called races, or subspecies. For example, we have seen that the species *E. burchelli* is composed of at least three rather distinct (but intergrading) races: Grant's zebra, Chapman's zebra, and Burchell's zebra. Races may be viewed as cohorts of local populations inhabitating different subdivisions of the geographic range of a species, and differing in one or more characteristics. A subspecies is simply a race that has been honored with a formal taxonomic name. As pointed out earlier, a species with recognized races is called polytypic, while those without are monotypic. Man is obviously polytypic. Polytypism is most likely found in forms with widely dispersed or disjunct populations that have become adapted to dissimilar habitats. When races meet in nature, there is a tendency toward the amalgamation of their distinctive features through hybridization. The amount of intergradation will depend primarily upon the extent of marginal contact.

Naming races is a purely subjective procedure based on two or more groups of populations attaining enough distinctions to warrant recognition of the fact. There is no objective rule to follow in determining when *enough* is found. Although the differences between races are objectively ascertainable facts, the number of races we choose to recognize is a matter of convenience, a cataloging device used to organize and record observed intraspecific diversity, which allows intelligible communication among students of species and evolution.

Any evolutionary event is highly contingent upon the extent and nature of the variation among members of a local population and differences between populations or races making up the species. Therefore, a complete description and classification of organic diversity is a necessary prerequisite for understanding the process of evolution. Also, there is great heuristic value in attempts to define the elementary categories of biological classification. We have gained some idea of the problems confronting us. In order to explain that theme of evolution designated *speciation*, it is necessary to investigate the factors involved in the origin of isolating mechanisms. These factors operate in the arena of the local interbreeding population, or deme, which will be examined from a genetic viewpoint in the following chapter.

References

Cain, A. J., *Animal Species and Their Evolution*. New York: Harper & Row, Publishers, Incorporated, 1954.

Ehrlich, Paul R., and Richard W. Holm, *The Process of Evolution*. New York: McGraw-Hill Book Company, 1963. See Chapter 13 for a less orthodox but thought-provoking viewpoint on species and some other basic concepts of evolution.

Grant, Verne, *The Origin of Adaptations*. New York: Columbia University Press, 1963. See Chapters 2, 4, and 12.

Heslop-Harrison J., *New Concepts in Flowering-Plant Taxonomy*. London: William Heinemann, Ltd., 1953. A small text on the impact of experimental and ecological studies on orthodox (typological) plant taxonomy.

Mayr, Ernst, "Speciation Phenomena in Birds," *Amer. Naturalist, 74* (1940), 249–78.

Mayr, Ernst, *Animal Species and Evolution*. Cambridge, Mass.: Harvard University Press, 1963. An authoritative treatment of the species category (see Chapters 1, 2, 3, 12, 13, and 14).

Simpson, George Gaylord, *Principles of Animal Taxonomy*. New York: Columbia University Press, 1961.

Three

Genes in Demes

In its essence, evolution involves changes in the genetic makeup of populations. An understanding of population genetics, therefore, is needed to fully appreciate the process.

Prediction of the frequency of genotypes among the offspring of a specified mating is possible, but exact prediction depends upon a knowledge of a great many factors: the loci and alleles concerned, the genotypes of the parents, the relative viability of the gametes, and the relative viability of the genotypes. Phenotypic ratios may be the same as, or quite different from, the frequencies of the genotypes because of additional factors such as dominance, epistasis, and penetrance. Mendelian genetics is primarily the study of these factors in order to understand the mechanisms by which the heritable material is transmitted from parent to offspring. Population genetics also deals with predicting the frequencies of genotypes and phenotypes from one generation to the next but applies the Mendelian concepts to all matings in an interbreeding population. Again, many factors must be known before a prediction can be made.

It is the purpose of this chapter to introduce the basic principles involved in extending Mendelian genetics to the level of populations. We shall show that predictions of genotypic and phenotypic frequencies may be readily calculated. The information required for such predictions, and the pitfalls and limitations in the methodology, are pointed out along the way.

Populations and demes

Few individuals exist alone in nature. Together they assume spatial-temporal relationships; they tend to aggregate. The platitudinous reason for this is simply that organisms become adapted (or suited) to particular sets of environmental conditions, and those similarly adapted are more apt to cluster in one kind of environment. The term *population*, when broadly defined, refers to any set of items, but it is used by biologists to describe specifically such aggregates of similarly adapted individuals. Population geneticists limit the term even more to describe groups of sexual forms that associate for reproduction as well as for ecological reasons. Genetically defined, a population (or more correctly, a genetic population) is a spatial-temporal group of conspecific interbreeding individuals. The genetic population maintains a continuity over time because of reproductive interconnections between generations, and is endowed with spatial unity owing to interbreeding among its members. A population may grow in size or become reduced through migration of individuals in or out, or by alterations in birth and death rates. It may fuse with other populations and it may become extinct, either by the total elimination or by the complete emigration of its members.

In this description of population, gene exchange between the members is the major consideration. In fact, many basic discoveries are made in population genetics by considering populations of genes rather than populations of individuals. In this light, all genetic information distributed among an interbreeding group of individuals collectively form a *gene pool*, which is temporarily dispersed and held as a set of particular genotypes. The zygotes (genotypes) that develop into individuals of one generation result from the union of gametes produced by the preceding generation. Alleles are sorted, shuffled, and then combined into gametes by the process of meiosis. The gametes combine during fertilization to produce a new set of genotypes. Consequently, a reconstituted gene pool is produced with each generation.

The gene pool concept is best represented by small isolated or semi-isolated colonies whose members can interbreed in a manner approximating panmixia, that is, mating pairs are a random association of genotypes. The terms *gamodeme* (Gilmour and Gregor), *panmictic unit* (Wright), and *local Mendelian population* (Dobzhansky) have been applied to such populations, although genetic population, or simply *deme*, is becoming the common designation.

Genotype frequencies in populations

The obvious starting point for the study of any population is an examination of the phenotypes of the individuals. Some variation always exists among the members of a population. A quick glance at the characteristics of one's codemic acquaintances will readily confirm this fact. To a certain degree, the observed phenotypic variation is based on underlying genetic diversity. To describe the population in genetic terms, we must next determine the genetic nature of given traits by analyzing the results of prescribed matings. Once the mode of inheritance of any trait is known, we can then treat the population in terms of the relative number of genotypes in the adult (diploid) phase and the relative number of alleles in the gametic (haploid) phase.

Consider the simple case of a single autosomal locus, A, with either of two alleles, A or a, possible at this locus. The number of individuals of each of the three genotypes, AA, Aa, and aa, is equivalent to the actual number of individuals in the three phenotypic classes if dominance is lacking, i.e., if each genotype is phenotypically distinguishable. Let the algebraic expressions n_1, n_2, and n_3 represent the numbers of the genotypes AA, Aa, and aa, respectively, where $n_1 + n_2 + n_3 = N$. In this hypothetical case, N signifies the total number of individuals in the population. When actually analyzing large populations and when it is impractical to tabulate each organism, we let N become the number of individuals in a sample. How representative of the population the sample actually is depends essentially upon the technique of the investigator.

Symbolically, the proportions, or percentages, of AA, Aa, and aa can be designated x, y, and z, respectively, and can be referred to as genotype frequencies, where

$$\frac{n_1}{N} = x \qquad \text{(frequency of } AA\text{)}$$

$$\frac{n_2}{N} = y \qquad \text{(frequency of } Aa\text{)}$$

$$\frac{n_3}{N} = z \qquad \text{(frequency of } aa\text{)}$$

The sum of x, y, and z is equal to unity, or 100 percent.

Gene frequencies in populations

The total number of diploid individuals in the population N was constructed from $2N$ gametes donated by the preceding generation. A certain number of these gametes possessed the A allele; the remainder carried the a allele. Different combinations of A-allele gametes and a-allele gametes formed the diploid population by uniting during fertilization. The number of AA genotypes, designated as n_1, was produced by the union of twice that number of gametes, $2n_1$, each possessing a single A allele. The heterozygotes were formed from $2n_2$ gametes, n_2 possessing the A allele and n_2 having the a allele. The total number of A-allele gametes involved in producing diploid individuals equals the sum of $2n_1$ and n_2. Their proportion (symbolized by p) among all gametes is

$$p = \frac{2n_1 + n_2}{2N} = \frac{n_1 + \frac{1}{2}n_2}{N}$$

The proportion of gametes with the a allele (represented by q) can be determined by considering the number of a-allele gametes which formed the Aa and aa individuals. Consequently

$$q = \frac{n_2 + 2n_3}{2N} = \frac{\frac{1}{2}n_2 + n_3}{N}$$

Such proportions are called gene frequencies.

Since these are considered the only alleles at the locus, $p + q = 1$.

Thus
$$p + q = \frac{n_1 + \frac{1}{2}n_2}{N} + \frac{\frac{1}{2}n_2 + n_3}{N}$$

$$= \frac{n_1 + n_2 + n_3}{N}$$

$$= 1$$

and
$$p = 1 - q$$

Gene frequencies can be determined from the genotype frequencies, since

$$p = \frac{n_1 + \frac{1}{2}n_2}{N} = \frac{n_1}{N} + \frac{\frac{1}{2}n_2}{N} = x + \frac{1}{2}y \qquad \text{(Eq. 3.1)}$$

and
$$q = \frac{\frac{1}{2}n_2 + n_3}{N} = \frac{\frac{1}{2}n_2}{N} + \frac{n_3}{N} = \frac{1}{2}y + z \qquad \text{(Eq. 3.2)}$$

Simply stated, the gene frequency of the allele A is determined by adding half of the percentage of the heterozygotes to the percentage of the AA homozygotes, and the frequency of the allele a is determined by adding half of the heterozygote class to the proportion of aa homozygotes.

The M-N blood types of man can be used to illustrate the calculation of gene frequencies. Family data have shown that the M-N blood types of man are determined by a pair of alleles, L^M and L^N, at a single autosomal locus. The three blood types M, MN, and N, are clearly distinguished and correspond to the genotypes $L^M L^M$, $L^M L^N$, and $L^N L^N$. The numbers and proportions of these genotypes in samples from three populations are present in Table 3.1. The gene frequencies of alleles L^M and L^N can be estimated for each of these populations by using either the numbers of individuals in the samples or the genotype frequencies (Egs. 3.1 and 3.2). For example, the gene frequencies of L^M and L^N in the sample of 730 Australian aborigines are

$$p \text{ (frequency of } L^M) = \frac{22 + \frac{1}{2}(216)}{730} = 0.0301 + \frac{1}{2}(0.2959) = 0.178$$

Table 3.1. Numbers and frequencies of M-N blood types in three human populations*

Population	Number in sample	Number and proportion of blood types			Allele frequencies	
		M	MN	N	$L^M(p)$	$L^N(q)$
Whites (United States)	6,129	No. 1,787 % 29.16	3,039 49.58	1,303 21.26	0.5395	0.4605
Navaho Indians (United States)	361	No. 305 % 84.49	52 14.40	4 1.11	0.917	0.083
Aborigines (Australia)	730	No. 22 % 3.01	216 29.59	492 67.40	0.178	0.822

* Data from various sources.

$$q \text{ (frequency of } L^N) = \frac{\frac{1}{2}(216) + 492}{730} = \frac{1}{2}(0.2959) + 0.6740 = 0.822$$

The equilibrium population

So far we have been describing a particular generation of a population. What will the gene and genotype frequencies be in the subsequent generations? The genotypes in the next immediate generation, or progeny, are constructed from a sample of the gametes produced by the parents. The kinds of gametes in the sample are determined by several factors that influence the process by which genes are transmitted from one generation to the next. These agencies must be known before the above question can be answered. Although some of them will be discussed in greater detail later, we shall introduce the factors here in order to make clear that the conclusions reached in this discussion are based on several important assumptions.

Mode of inheritance

We shall be considering, primarily, sexually reproducing organisms consisting of diploid adults with haploid gametes. As mentioned earlier, the site of the locus (autosomal or sex-linked), the number of alleles, the number of loci, and the dominance and epistasis relationships must be determined before any calculations of the progeny generation can be made.

Population size

Sampling variations due to chance alone can be significant in determining what genes are present among the sample of gametes that form the progeny generation. All populations are finite, and the smaller the size, the greater the chance that the gene frequencies in the gametes will be different from those in the parental population owing to sampling error. For the present we shall consider only "large" populations, in which random changes in gene frequencies are insignificant enough to be ignored.

Mating systems

The mating system most common in obligately outcrossing populations approximates random breeding (panmictic units), in which an individual has an equal probability of mating with any member of the opposite sex, regardless of its genotype. This is not quite true of a large population occupying an extensive range, because an individual usually selects a mate from the members of a proximate group. This restriction is not significant unless there is a propensity toward extensive ancestral relationship between partners (inbreeding), which occurs if a population subgroup is small and partially isolated or if, as in many plant groups, self-fertilization occurs. The essential point is that the phenotypic expressions of the alleles under consideration are not among those that form the criteria for selecting mates (assortative mating).

Mutation and migration

Mutations can change one allelic state to another. The gene frequencies among the gametes may not reflect those of the parents if mutations occur, unless the occurrence is equal in both directions, that is, the number of forward mutations (A mutating to a) equals the number of reverse mutations (a mutating to A).

Likewise, the gene frequencies can be so modified by the introduction or withdrawal of genotypes in disproportionate ratios that they alter the parental generation. The immigration of foreign individuals into the population or the emigration of members out of the population changes the frequencies of the gametes being produced (alters the constitution of the parental generation) when the migrants have gene frequencies different from those of the original population.

Selection

Often the phenotypic expressions of the various genotypes include differences in fertility and viability. The differential ability of the parental genotypes to contribute functional gametes and the differential survival rates of the various gametes and/or zygotes would definitely change gene and genotype frequencies of the offspring compared with those of the parents. The unequal transmission of genes to subsequent generations by different genotypes constitutes selection, which is the subject of Chapter 5. It is impossible to ignore selective forces in comparing actual gene frequencies of different generations of a population, but let us do so in the following hypothetical case in order to keep the discussion as uncomplicated as possible.

We shall now answer the question asked above: What are the gene and genotype frequencies in the progeny generation? First consider the hypothetical case of alternate alleles at the single autosomal locus A with the conditions that the genotypes form a large, random-mating, and closed (no migration) population within which there is no mutation or selection. In short, we assume at the outset that no factor operates which would cause the gene frequencies in the offspring generation to differ from those in the parental population.

The clear but cumbersome approach to the problem is to consider (1) the kinds of matings in the population, (2) how frequently each mating occurs, (3) the ratios of the offspring genotypes expected for each kind of mating, and (4) the combined ratios of the three genotypes in the progeny generation, considering all matings.

The possible mating types are found by making female-male permutations of all the genotypes in the population; there are nine such combinations possible with the three genotypes AA, Aa, and aa.

With random mating, the frequency of each type of mating will depend upon the frequencies of the genotypes. When an individual chooses a mate randomly, the likelihood of the partner's being a particular genotype equals the frequency of that genotype. Thus, the proportion of each mating combination equals the product of the frequencies of the genotypes. For example, the $AA \times AA$ mating occurs with the frequency x^2 among all matings possible in the population. Here we must specify that similar genotypes occur with the same frequencies in the two sexes ($x_♀ = x_♂$). The nine kinds of matings and their frequencies are as shown in the table.

	Males		
	AA	Aa	aa
Females	x	y	z
AA	$AA \times AA$	$AA \times Aa$	$AA \times aa$
x	x^2	xy	xz
Aa	$Aa \times AA$	$Aa \times Aa$	$Aa \times aa$
y	xy	y^2	yz
aa	$aa \times AA$	$aa \times Aa$	$aa \times aa$
z	xz	yz	z^2

The three genotypes AA, Aa, and aa will be represented among the genotypes produced by the matings described above. However, the genotypic proportions of the offspring may be different from those in the parental generation. Since selective forces are assumed to be inoperative, the average number of offspring produced by each mating is considered to be the same for all, regardless of the genotypes in the cross. Thus, the proportion of the progeny produced by each of the nine mating classes can be considered equivalent to the frequency of each mating class. For example, the $AA \times AA$ cross occurs with a frequency of x^2. The offspring of these particular matings equal x^2 of all progeny produced by the total population. Each of the progeny from these matings will have the genotype AA. The $Aa \times Aa$ matings occur with a frequency of y^2. Of the progeny produced by matings of this type, $\frac{1}{4}y^2$, $\frac{1}{2}y^2$, and $\frac{1}{4}y^2$ will be AA, Aa, and aa, respectively. By following this procedure for all matings, we can arrive at the total frequency of each progeny genotype. The nine mating types, their frequencies, and the proportion of offspring produced by each mating are summarized in Table 3.2.

Table 3.2.   Matings and offspring in a random-mating population

Type of mating	Frequency of mating	Offspring		
		AA	Aa	aa
$AA \times AA$	x^2	x^2		
$AA \times Aa$				
$Aa \times AA$	$2xy$*	xy	xy	
$AA \times aa$				
$aa \times AA$	$2xz$*		$2xz$	
$Aa \times Aa$	y^2	$\frac{1}{4}y^2$	$\frac{1}{2}y^2$	$\frac{1}{4}y^2$
$Aa \times aa$				
$aa \times Aa$	$2yz$*		yz	yz
$aa \times aa$	z^2			z^2
Totals	$(x+y+z)^2$ $=1$	$(x+\frac{1}{2}y)^2$ $=p^2$	$2(x+\frac{1}{2}y)(\frac{1}{2}y+z)$ $=2pq$	$(\frac{1}{2}y+z)^2$ $=q^2$

* Reciprocal classes are combined and their frequencies are added.

The frequency of AA offspring produced by the four matings $AA \times AA$, $AA \times Aa$, $Aa \times AA$, and $Aa \times Aa$ equals $x^2 + xy + \frac{1}{4}y^2$, which factors to $(x + \frac{1}{2}y)^2$. Since $x + \frac{1}{2}y$ is equal to p (Eq. 3.1), the frequency of the AA genotype in the progeny generation can be expressed as the square of the frequency of the A allele, p^2.

The heterozygous progeny Aa are produced by seven types of matings and occur with the total frequency of $xy + 2xz + \frac{1}{2}y^2 + yz$, which factors to $2(x + \frac{1}{2}y)(\frac{1}{2}y + z)$, which in turn equals $2pq$. Thus the genotype frequency of Aa in the progeny generation equals twice the product of the two gene frequencies. Finally, by summing the frequencies of aa offspring produced by the matings $Aa \times Aa$, $Aa \times aa$, $aa \times Aa$, and $aa \times aa$, we obtain $(\frac{1}{2}y + x)^2 = q^2$. The genotype frequencies x, y, and z in the original parental population become p^2, $2pq$, and q^2 after one generation of random mating.

Since we know the genotype frequencies, we can now determine the gene frequencies in the progeny generation. As stated earlier, the proportion of the A alleles equals half of the heterozygote frequency added to the proportion of the AA homozygotes. Letting p_1 stand for the gene frequency of A in the progeny generation,

$$p_1 = p^2 + \frac{1}{2}(2pq)$$
$$= p^2 + pq$$
$$= p(p + q)$$

and since $p + q = 1$, we find that $p_1 = p$. The gene frequency of A in the progeny is the same as that which originally existed in the parental population! Of course the same is true for the gene frequency of a. This should not be too surprising since we specifically omitted all the factors which would cause a change in gene frequencies in our theoretical model.

On the other hand, the genotype frequencies x, y, and z become p^2, $2pq$, and q^2 in a single generation of random mating. The population $p^2 : 2pq : q^2$ is considered an equilibrium population, since the gene and genotype frequencies remain the same in each succeeding generation. The reader may demonstrate this by using the method shown above, substituting p^2 for x, $2pq$ for y, and q^2 for z.

Let us now examine an alternative and much simpler method for determining genotype frequencies in succeeding generations. If, as was suggested above, mates are selected at random and gamete union in each mating combination is also random, results equivalent to those above could be obtained by completely ignoring the mating combinations and considering simply the random union of *all* the gametes produced by the parental population. Such a "pooling" and random uniting of gametes is approximated in some aquatic forms through external fertilization. Populations of other organisms can be envisioned in the same light as possessing a gene pool when mating is random.

The composition of the progeny generation can be ascertained simply by determining the kinds and frequencies of the gametes produced by the preceding generation and calculating the frequencies of the genotypic combinations formed by random fertilization. Let us say that all males contribute equally to a sperm gamete pool and that all females contribute equally to an egg gamete pool. We can view each pool as containing p gametes possessing the A allele and q gametes possessing the a allele. The frequency of AA, Aa, and aa zygotes, produced by the union of sperm and eggs drawn at random from the two gamete pools, would be p^2, $2pq$, and q^2. These proportions are obtained by multiplying the frequencies of the different gametes in the sperm pool by those in the egg pool, $(p_\female + q_\female)(p_\male + q_\male)$. The genotype frequencies of the progeny are the square of the gene frequencies among the gametes when $p_\female = p_\male$ and $q_\female = q_\male$. Thus

$$(p + q)^2 = p^2 + 2pq + q^2 \qquad \text{(Eq. 3.3)}$$

In order to visualize this idea more clearly, the relationship of the gametic gene frequencies $p = 0.6$ and $q = 0.4$ and the resulting genotype frequencies are presented diagrammatically in Fig. 3.1. In this

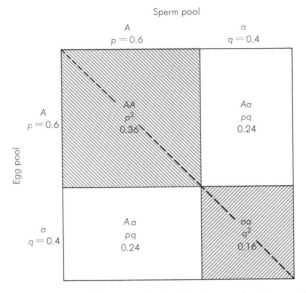

Fig. 3.1. An equilibrium population. The diagram depicts an equilibrium population resulting from the random union of gamete pools. Each of the four lines forming the square is set equal to unity. The top and left sides of the square are each divided into two sections representing the respective gene frequencies in the male and female gamete pools ($p = 0.6$ and $q = 0.4$). The square is divided into four areas by a vertical line and a horizontal line, originating at the dividing points at the top and left sides, respectively. These areas depict the relative genotype proportions of the offspring formed by randomly uniting the sperm and egg gamete pools. The total area of the square equals 1.0 and, thus the sum of the "genotype" areas equals 1.0. The vertical and horizontal lines intersect at a point along a diagonal line drawn from the upper left corner to the lower right corner. The point of intersection, referred to as the *population point*, divides the diagonal line into two sections proportional to the gene frequencies in the gamete pools.

case, the offspring produced by the random union of gametes are 0.36 *AA*, 0.48 *Aa*, and 0.16 *aa*. These genotypic proportions are said to be in equilibrium since they would be the same in future generations. Any value assigned to q (with p understood to be $1 - q$) would lead to such an equilibrium under the conditions originally set forth. For example, when $q = 1.0$, the genotypic ratios are $p^2 = 0$; $2pq = 0$; $q^2 = 1.0$. Here the equilibrium is considered to be a trivial case, because the total population would consist of only homozygous *aa* individuals. When the gene frequencies are equal, $p = q = \frac{1}{2}$, the square in Fig. 3.1 would be divided into four equal areas, each one 25 percent of the total. Among all possible equilibrium populations, this particular one is outstanding since it possesses the maximum percentage of heterozygotes, 50 percent (Fig. 3.2).

Populations with identical gene frequencies do not necessarily have similar genotype frequencies. The following four populations have gene frequencies of 0.6 *A* and 0.4 *a*:

Fig. 3.2. Genotype frequencies represented by a single *point* in an equilateral triangle. The height of the triangle is set at one, and perpendicular lines from point P (called the population point) to the three sides equal the proportions *x*, *y*, and *z* (or p^2, $2pq$, and q^2). The projection point Y divides the base into segments XY and YZ, which are proportional to the gene frequencies *p* and *q*, respectively. In the lower figure, the four populations (I, II, and III, and IV) all have the same gene frequencies ($q = 0.4$); however, only population II is in Hardy-Weinberg equilibrium. Its population point lies on the parabola, which represents the locus of all equilibrium populations. Note that the vertex point of the parabola (represented by the open circle) depicts that particular equilibrium population in which $p = q = 0.5$, and in which the proportion of heterozygotes, $2pq$, is maximal (0.5).

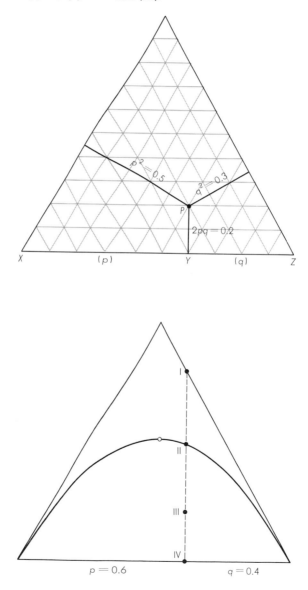

	AA	Aa	aa	p	q
I	0.20	0.80	0	0.6	0.4
II	0.36	0.48	0.16	0.6	0.4
III	0.50	0.20	0.30	0.6	0.4
IV	0.60	0	0.40	0.6	0.4

Although they have different genotype frequencies, each of the four populations will reach (or, in the case of population II, remain at) the same equilibrium state of 0.36 AA : 0.48 Aa : 0.16 aa after a single generation of random mating under the conditions previously set forth (Fig. 3.2).

Populations in equilibrium will continue to produce similar genotype ratios from generation to generation. This concept of the constancy of genotypic ratios in random-mating populations, when gene frequencies are specified as constant, is referred to as the Hardy-Weinberg law in honor of the two men who independently formulated the idea in 1908. It has served as the cornerstone for the development of population genetics.

Testing population samples

Let us return to the example of the M-N blood types in man (Table 3.1) and determine whether the proportions observed in the sample of Australian aborigines are those to be expected in a random-mating equilibrium population according to the Hardy-Weinberg principle. If the 730 individuals are a representative sample and if the population is actually in equilibrium, then the estimated gene frequencies ($p = 0.178$, $q = 0.822$) should represent the true proportions in the gene pool. Furthermore, the *expected* genotype frequencies, determined by squaring the gene frequencies under the Hardy-Weinberg principle, are

$$(0.178L^M + 0.822L^N)^2 = 0.0317L^ML^M + 0.2926L^ML^M + 0.6757L^NL^N$$

The numbers observed in the sample of 730 individuals and those expected if we assume random mating are

	Numbers observed	Numbers expected
L^ML^M	22	23.1
L^ML^N	216	213.6
L^NL^N	492	493.3
Total	730	730.0

Since the gene frequencies were estimated from a sample, they are subject to sampling error, that is, they may not represent the true values in the total population. These estimates were used in determining the number of each genotype under the hypothesis of a Hardy-Weinberg equilibrium. Usually the Chi-square method is used to ascertain the extent of agreement between the numbers observed in the sample and those expected according to the hypothesis. In this case the genotype numbers observed and those expected have a Chi-square value of 0.083 with one degree of freedom (76 percent probability), indicating very close agreement. Therefore it is suggested that the individuals in the population are indeed mating at random (are panmictic) with respect to M-N blood groups. This seems reasonable, since few human beings know to which M-N blood group they belong, and those who do know would seldom include the trait among their marital qualifications. If the numbers are dissimilar, then the lack of panmixia and/or any other of the previously listed factors would be involved in producing the discrepancy.

Some students may think that the expected numbers should *always* reflect closely the observed numbers, since the former are estimated from gene frequencies determined from the latter. To confirm that this is not always true, we can run the test on a hypothetical "population" known *not* to have arisen by random mating. For example, let us consider the combined data on Australian aborigines and Navaho Indians as having been obtained from a single hypothetical population.

	M	M-N	N	*Total*
Australian aborigines	22	216	492	730
Navaho Indians	305	52	4	361
Hypothetical population	327	268	496	1,091

What are the gene frequencies in the artificial population? What are the genotype frequencies and numbers expected in the population, assuming random mating and the other conditions of the Hardy-Weinberg model? A Chi-square test of these values would show that it is rather unlikely that the deviations are due to chance. However, it should be realized that the test does not specify what factor (nonrandom mating, selection, migration, etc.) is responsible for the lack of agreement.

In addition to conforming well to the conditions of the Hardy-Weinberg model, the M-N blood groups are excellent for demonstrating an equilibrium population, because the heterozygotes MN are differentiable and can be included in the data for estimating gene frequencies. If the trait being considered is controlled by a recessive gene, on the other hand, the heterozygotes resemble the dominant homozygotes and together form a single phenotypic class. The estimation of these gene frequencies can be made only by using the proportion of recessive homozygotes and presupposing genetic equilibrium ($z = q^2$). For instance, recessive feeblemindedness (phenylketonuria) occurs in about 1 in 40,000 human beings. If equilibrium prevails, the frequency of affected individuals, 1/40,000, should equal q^2, and the percentage of the recessive alleles, q, is the square root of the genotype frequency, or 1/200. The frequency of the dominant allele is $1 - q$, or $1 - 1/200 = 199/200$. The equilibrium genotypic proportions can now be calculated by expanding the gene-frequency binomial $(p + q)^2$.

$$p^2 = \left(\frac{199}{200}\right)^2 \qquad = \frac{39,601}{40,000}$$

$$2pq = 2\left(\frac{199}{200}\right)\left(\frac{1}{200}\right) = \frac{398}{40,000}$$

$$q^2 = \left(\frac{1}{200}\right)^2 \qquad = \frac{1}{40,000}$$

Fig. 3.3. An example of an equilibrium population in which one allele is rare and is carried primarily in heterozygotes. In this case, $q = 0.05$, and heterozygotes are 38 times as frequent as aa homozygotes.

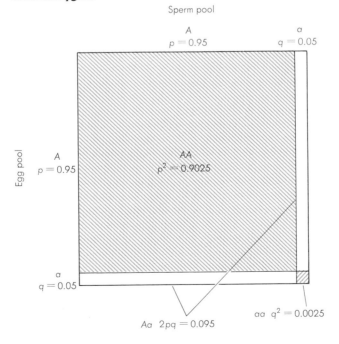

Sperm pool

A $p = 0.95$ a $q = 0.05$

Egg pool

A $p = 0.95$

AA $p^2 = 0.9025$

a $q = 0.05$

aa $q^2 = 0.0025$

Aa $2pq = 0.095$

A direct comparison of expected and observed proportions cannot be made and random mating cannot be tested since, in order to procure gene frequencies, equilibrium is assumed in the first place.

Another fact can be demonstrated with this example: when an allele is rare, the majority of individuals possessing the alleles are heterozygotes (Fig. 3.3). There are 398 times as many expected carriers (heterozygotes) of the phenylketonuria allele as affected (homozygous recessive) individuals. The ratio of carriers to those affected is even greater for those recessive traits which occur less frequently than phenylketonuria. For example, recessive alcaptonuria, which is a disorder of amino-acid metabolism, occurs in about 1 in 1 million people. The frequency of the recessive allele is approximately 1/1,000, and the ratio of heterozygotes to recessive homozygotes is 1,998 to 1. This clearly emphasizes the fact that rare alleles occur primarily in the heterozygous state. It follows that the majority of individuals possessing a rare recessive trait have normal, though heterozygous, parents.

Multiple alleles

The principles of the Hardy-Weinberg equilibrium are not complicated by considering multiple alleles and letting all other conditions be as in the previous discussions. Consider three alleles, A_1, A_2, and A_3, with frequencies of p, q, and r, respectively, where $p + q + r = 1$. The genotypes A_1A_1, A_1A_2, A_1A_3, A_2A_2, A_2A_3, and A_3A_3 will exist in a population in equilibrium in ratios of $p^2(A_1A_1) + 2pq(A_1A_2) + 2pr(A_1A_3) + q^2(A_2A_2) + 2qr(A_2A_3) + r^2(A_3A_3)$, which is the square of the array of the gametic gene frequencies, $(p + q + r)^2$. The principle of establishing equilibrium in one generation for nonequilibrium populations holds for multiple alleles just as it does for two alleles. An example involving three alleles ($p = 0.3$, $q = 0.5$, and $r = 0.2$) is presented in Fig. 3.4. Again, the equilibrium state is established in one generation through random mating and random union of gametes.

Gene frequencies of multiple alleles can be estimated easily from genotype proportions when all genotypes are phenotypically distinguishable. The frequency of the homozygotes of any one of the alleles is added to half of the proportion of each class of heterozygotes involving that allele. Thus, in the case of three multiple alleles,

$$p = p^2 + pq + pr \qquad q = q^2 + pq + qr \qquad r = r^2 + pr + qr$$

Sometimes only one of a series of alleles is of particular interest. It is possible to treat a multiple group as a pair of alleles by considering the frequency of one of them, say A_2 (represented by q), and letting the total frequency of all the others be $1 - q$. The Hardy-Weinberg binomial then becomes

$$[q + (1 - q)]^2 = q^2 + 2q(1 - q) + (1 - q)^2$$

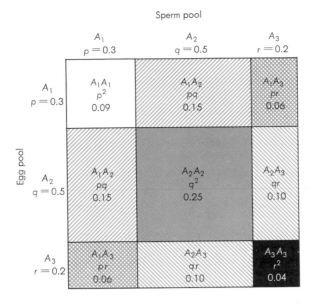

Fig. 3.4. Multiple alleles in an equilibrium population. The three alleles, A_1, A_2, and A_3, occur with frequencies $p = 0.3$, $q = 0.5$, and $r = 0.2$ in each gamete pool. The genotype frequencies are given as the areas within the square, which are formed by geometrically squaring the gametic array.

Equilibrium populations and sex

Similarity of gene frequencies in the two sexes is an assumption of the Hardy-Weinberg law. Equilibrium is reached after two generations when a population consists of males and females with dissimilar gene frequencies: allelic proportions in the sexes become equalized in one generation and genotypes reach equilibrium in the following. This can be seen diagrammatically in Fig. 3.5. The sperm pool consists of 80 percent A-allele gametes, while only 40 percent of the eggs are A-allele gametes. The vertical and horizontal lines separating the square into sections (representing the first-generation progeny) do not meet on the diagonal line. The population is not in equilibrium. Equilibrium will be reached with $p = 0.6$ in the next generation. This value can be derived from the illustration by taking the midpoint of that section of the diagonal line between the two points crossed by the vertical and horizontal lines. Simply stated, the equilibrium frequency of an allele is the average of the frequencies of that allele found in both sexes.

When the alleles are carried on the sex chromosomes, the locus is sex linked. We shall consider the case of homogametic (XX) females and heterogametic (XY) males. The alleles in the population are distributed unequally, since they are carried on chromosomes which occur disproportionately in the sexes. Two-thirds are carried by the females and one-third by the males. Again consider only two alleles, A and a, with these possible genotypes and frequencies:

	Females			*Males*	
Genotypes	AA	Aa	aa	A	a
Frequencies	x	y	z	m	n

When determining genotype and gene frequencies of sex-linked alleles, each sex is treated as a separate subpopulation. Gene frequencies among the females are determined in the same manner as those for an autosomal gene, that is, $p_♀ = x + \tfrac{1}{2}y$, whereas in the males the gene frequencies are identical with the genotype frequencies $p_♂ = m$ and $q_♂ = n$. The population is in equilibrium, or will reach equilibrium in a single generation of panmixia, when $q_♀ = q_♂$. If the population consists of equal numbers of males and females, the gene frequency of the a allele for the total population q_T is

$$q_T = \tfrac{2}{3}q_♀ + \tfrac{1}{3}q_♂$$

When $q_♀$ and $q_♂$ are unequal, the population is not in equilibrium.

Fig. 3.5. A population formed when the gene frequencies between the sexes are dissimilar. If a diagonal line is drawn, starting at the upper left-hand corner, we find that the horizontal and vertical lines do not intersect at a place along the diagonal line, because the gene frequencies are not the same in the two gamete pools ($q_♂ \neq q_♀$). The *population point* will lie on the diagonal when equilibrium is reached. The exact location can be ascertained by locating the midpoint of that segment of the diagonal formed where it is crossed by the horizontal and vertical lines. Only one-third the distance is used in the case of a sex-linked locus (depicted by the circle in this example).

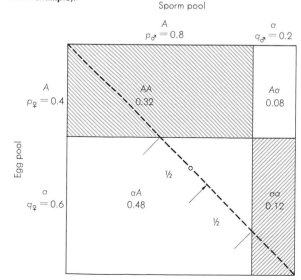

In this situation a stable condition is not reached in one generation of random mating, or even in two, but is rapidly approached over generations. The over-all value q_T does not change in each succeeding generation leading to equilibrium—the same principle is shown with an autosomal locus. On the other hand the individual gene frequencies of the two sexes ($q_♀$ and $q_♂$) oscillate from one generation to the next. This is true because $q_♂$ in each generation equals $q_♀$ in the prior generation and $q_♀$ of any one generation is equivalent to $\frac{1}{2} q_♀ + \frac{1}{2} q_♂$ in the one preceding it.

We can return to Fig. 3.5 for an illustration, but now it is necessary to think in terms of a sex-linked locus. The male gamete pool is $p_♂ = 0.8$ and $q_♂ = 0.2$, while the egg pool has $p_♀ = 0.4$ and $q_♀ = 0.6$. We find that $q_T = \frac{2}{3} (0.6) + \frac{1}{3} (0.2) = 0.467$ and that $p_T = 0.533$.

The diagram depicts the random union of the gamete pools. Now the areas within the square represent only the genotype frequencies of the females ($AA_♀ = 0.32$, $Aa_♀ = 0.56$, and $aa_♀ = 0.12$). The proportion of a alleles among the female progeny (symbolized by $q'_♀$) is $q'_♀ = 0.12 + \frac{1}{2}(0.56) = 0.4$. The genotype frequencies of the male offspring m and n can be represented by the two sections of the line forming the right side of the square. The two segments are also equivalent to the gene frequencies of the male offspring ($p'_♂$ and $q'_♂$). Obviously the genotype frequencies in the male offspring ($p'_♂$ and $q'_♂$) are the same as the gene frequencies among their mothers ($p_♀$ and $q_♀$), as seen by the fact that the segments of each side of the square are equivalent.

Let us review the consequences of one generation of random mating in this example. Only one frequency, q, will be considered in the following discussion, since p is understood to be $1 - q$.

Parental generation	*Progeny generation*
$q_♀ = 0.6$	$q'_♀ = 0.4$
$q_♂ = 0.2$	$q'_♂ = 0.6$
$q_T = 0.467$	$q'_T = 0.467$

The total gene frequency does not change, so that $q_T = q'_T$. The equilibrium value of the a-allele frequency, $\hat{q}$ (read "hat" q), will also be the same as that of the original value, q_T. The equilibrium value can be found in Fig. 3.5 by locating a point a third of the distance from the horizontal line on that segment of the diagonal between the points where it is crossed by the vertical and horizontal lines (this point is represented by a circle in the figure).

In the parental generation, $q_♂$ is less than the equilibrium value (0.467), but $q'_♂$ is greater than this. The reverse is true for the female proportion ($q_♀ > 0.467 > q'_♀$). In succeeding generations the gene frequencies of both sexes alternate above and below the over-all gene frequency, which remains constant. This is the oscillating nature of gene-frequency change in the two sexes mentioned above.

The difference between q'_σ and q'_φ among the offspring is half of the amount between q_σ and q_φ in the parental generation. In fact, the difference is halved each generation, approaching zero at equilibrium.

Multiple loci and genetic equilibrium

All autosomal loci considered separately attain an equilibrium of genotype frequencies in a single generation of random mating. But when two or more loci are considered jointly, the multifactorial genotypes such as *AABB*, *AABb*, *AAbb*, and *AaBB*, gradually approach genetic equilibrium over many generations rather than in one. A simple illustration of this point will suffice. Consider alternate alleles segregating at each of two independently assorting loci (*A, a* and *B, b*), where the gene frequencies of the respective alleles are p (*A*), q (*a*), r (*B*), and s (*b*). The monofactorial genotype frequencies at equilibrium are p^2 (*AA*) : $2pq$ (*Aa*) : q^2 (*aa*) for locus *A* and r^2 (*BB*) : $2rs$ (*Bb*) : s^2 (*bb*) for locus *B*; considered together, the proportions of the bifactorial genotypes are $(p^2 + 2pq + q^2)(r^2 + 2rs + s^2)$, which is equivalent to squaring the frequencies of the *gametes AB, Ab, aB,* and *ab*. If equilibrium prevails, the frequency of each sort of gamete is given as the product of the frequencies of the alleles contained therein. Thus, if we denote the gametic frequencies,

Gamete type	*AB*	*Ab*	*aB*	*ab*
Frequency	h	i	j	k

then at equilibrium, $h = pr$, $i = ps$, $j = qr$, and $k = qs$. In addition we find at equilibrium that $hk = ij$.

For a numerical example, consider a population with the following *gametic* frequencies:

Gamete type	*AB*	*Ab*	*aB*	*ab*
Frequency	$h = 0.40$	$i = 0.30$	$j = 0.20$	$k = 0.10$

The *gene* frequencies in this case are $p = 0.7$, $q = 0.3$, $r = 0.6$, and $s = 0.4$. This population is not in equilibrium with respect to these alleles, since the gametic frequencies do not equal those expected from the estimated gene frequencies, nor does $hk = ij$. The difference (d), $hk - ij$, with respect to two unlinked loci is halved each generation of random mating, until at equilibrium $hk - ij = 0$. When the population reaches equilibrium, the gamete frequencies will be

Gamete type	*AB*	*Ab*	*aB*	*ab*
Frequency	$h = pr$	$i = ps$	$j = qr$	$i = qs$
	$= 0.42$	$= 0.28$	$= 0.18$	$= 0.12$

To help clarify the idea of the attainment of equilibrium involving two loci, let us review the consequence of random mating in two simplified situations. First, consider a population which initially consists of *AaBb* heterozygotes only. The individuals will produce four kinds of gametes and, assuming the loci are unlinked, the different gametes are expected in equal frequencies (¼ *AB*, ¼ *Ab*, ¼ *aB*, and ¼ *ab*). The random union of the gametes will result in the nine genotypes possible for two loci with two alleles each; these genotypes are expected to occur in the familiar Mendelian F_2 dihybrid proportions. In the next generation the four types of gametes will again be produced in equal frequencies; and the random union of these will result in genotypes in the same proportions as those in the preceding generation. Therefore, equilibrium is reached in one generation of random mating. Now let us consider the second simple case, that involving a population made up of equal numbers of *AABB* and *aabb* individuals. Here only two kinds of gametes are produced (½ *AB* and ½ *ab*). These will unite at random to produce only three different genotypes,

Fig. 3.6. The approach to equilibrium of two loci, considered jointly. The difference (d) between the products of the coupling gamete frequencies and the repulsion gamete frequencies are graphed over successive generations, starting with equal frequencies of *AABB* and *aabb* individuals. The five curves refer to different degrees of recombination (R) between the two loci. Calculation of data courtesy of Henry Schaffer.

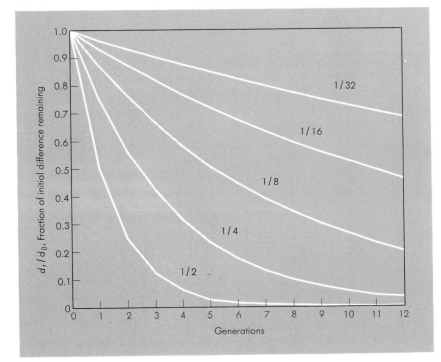

AABB, AaBb, and *aabb,* in a 1 : 2 : 1 ratio. In the next generation the gamete pool will consist of $\frac{3}{8}$ *AB*, $\frac{1}{8}$ *Ab*, $\frac{1}{8}$ *aB*, and $\frac{3}{8}$ *ab*. In the third generation the gamete proportions become $\frac{5}{16}$ *AB*, $\frac{3}{16}$ *Ab*, $\frac{3}{16}$ *aB*, and $\frac{5}{16}$ *ab*. There will be an increase in the frequency of the so-called *repulsion* gametes (that is, *Ab* and *aB*) and a proportionate decrease in the *coupling* gametes (that is, *AB* and *ab*) over generations until all gametes occur with equal frequencies, at which time equilibrium will have been attained. Since only coupling gametes are produced in the first generation of this population, *hk* and *ij* are not equal, and *d* is $\frac{1}{4}$. The value for *d* is $\frac{1}{8}$ in the second generation (that is, $\frac{3}{8} \times \frac{3}{8} - \frac{1}{8} \times \frac{1}{8} = \frac{1}{8}$), $\frac{1}{16}$ in the third, $\frac{1}{32}$ in the fourth, and continues to be halved each generation until equilibrium is reached.

Equilibrium is approached rather rapidly when the loci concerned are unlinked ($R = \frac{1}{2}$). But if the two loci are present on the same chromosome, the speed (number of generations) of equilibrium attainment is inversely proportional to the closeness of the linkage (Fig. 3.6).

An important consequence of this formulation is that at equilibrium, repulsion and coupling gametes are expected to be equally frequent, regardless of the degree of linkage. This means that the characters produced by alleles at linked loci show no particular association in equilibrium populations. Thus, when characters happen to associate in a population, the association may be due to alleles at separate loci which are in genetic disequilibrium resulting from recent population admixture. However, pleiotrophy, involving a single locus, might just as well explain the phenomenon. Like dominance, linkage can be detected only in prescribed breeding experiments at the family level and not directly from population data.

Genetic drift

A perfect equilibrium is reached when random breeding occurs in an ideally infinite population. This concept has been repeatedly stressed in the preceding sections. But at best only approximate states of equilibrium truly are reached, because any natural population, regardless how large, is a finite entity. Remember that the individuals of a particular generation are formed from $2N$ gametes of the preceding generation. Only when N is infinite can we be assured that the gametes perfectly represent the parental gene pool. The gametes carry only a *sample* of the parental gene pool and are subject to sampling error when N is finite. The smaller the population size and consequently the gametic sample, the greater the amplification of the sampling variation.

The gene frequency q may change in each succeeding generation. By chance it may increase or decrease from the value in the prior generation. Although the direction and the exact amount of the change is indeterminate, we can predict the odds, or probability, that it will be a certain value or lie within a particular set of values. The small

Greek letter delta (δ) is used to express the change in gene frequency due to such sampling variation. The change in the frequency of q is δq, where $\delta q = q_1 - q_0$ (the numerical subscripts represent generations). A theoretical frequency-distribution graph can be constructed for all values of δq possible for a trait in one generation in a finite population. The number of points in any such graph is $2N + 1$ and thus is limited only by the size of the population (the value of N). The magnitude of the change in gene frequency resulting from sampling error is expressed by the shape of the graph. The graph can be stated in statistical terms by the mean, or average, change in gene frequency ($\overline{\delta q}$) and the variance ($\sigma_{\delta q}^2$). The mean will be zero, because the change is just as likely to be positive as negative. The variance in one generation is

$$\sigma_{\delta q}^2 = \frac{p_0 \, q_0}{2N} \qquad \text{(Eq. 3.4)}$$

which is the square of the standard deviation (σ),

$$\sigma_{\delta q} = \sqrt{\frac{p_0 \, q_0}{2N}}$$

The variance increases as the population size N is reduced. The smaller the population, the greater the chance that the gene frequency will change by random-sampling variation.

Consider a panmictic population segregating for alleles A and a at an autosomal locus, with $p_0 = q_0 = 0.5$. Assume that the population size is limited to 50 individuals in the following generation. The mean of the possible values of q_1 is 0.5, since $\overline{\delta q} = 0$. The standard deviation is

$$\sigma_{\delta q} = \sqrt{\frac{0.5 \times 0.5}{2 \times 50}} = 0.05$$

The probabilities of the various possible q (frequency in progeny generation) values would approximate a *normal* distribution with a mean of 0.5. The probabilities of the occurrence of deviations in a normal distribution can be procured from available tables. In the tables we find that the chance of deviating beyond the range of one standard deviation (σ) on either side of the mean is approximately 32 percent. This means that there is a 32 percent chance that q_1 would be either greater than 0.55 or less than 0.45. Such random variations in gene frequencies continue in succeeding generations. They seem to wander, or drift with no apparent trajectory, toward a particular point. The ultimate fate of an allele is *fixation* ($q = 1$) or *loss* ($q = 0$). No further change is possible once the gene frequency reaches either zero or one; the individuals of the population are all homozygous for that locus (a homoallelic population), unless of course the lost allele is replaced by mutation or immigration. It is fortuitous which allele is fixed or lost. All loci tend toward homoallelism in very small populations, which results in decreased heterozygosity and loss of variability.

The value $p_0q_0/2N$ also expresses the variance of gene frequencies (either σ_q^2 or σ_p^2) after a single generation of panmixia among many populations of equal size and with similar initial gene frequencies. This is exactly the same concept as the variance of possible changes in a single population above, but now we are considering the variance of gene frequencies among several populations. Let us say that an experimenter establishes many laboratory populations, all of which have the same number of individuals and all of which possess similar initial gene frequencies. Each population will assume a new gene frequency, q_1, in the next generation. These new values will tend to differ, and there will be a dispersion of gene frequencies among the populations. The mean of the gene frequencies of all the populations should equal the original value, q_0. Here the variance is a measure of the amount of the divergence of gene frequencies among the populations (Fig. 3.7).

The variance increases each succeeding generation and at any given generation (t) is as follows (Crow, 1954):

$$\sigma_q^2 = p_0q_0 \left[1 - \left(1 - \frac{1}{2N} \right)^t \right] \qquad \text{(Eq. 3.5)}$$

Eventually, when t is very large, all the populations will become homoallelic and the variance will be maximal, $\sigma_q^2 = p_0q_0$. In other words, the a-allele frequency will be fixed ($q = 1$) in some of the populations and lost ($q = 0$) in the remaining portion. The relative proportion of the number fixed to the number lost is directly related to the proportion of p_0 and q_0. From Eq. 3.5 we see that the variance at a given time is dependent upon the size of the population, the number of generations that have elapsed, and the initial gene frequencies. These relationships can be seen in Table 3.3 for populations with $q_0 = 0.5$.

Table 3.3. The comparison of the variances over generations for populations of different sizes, all with $p_0 = q_0 = 0.5$*

Number of generations	*Size of population*					
	6	*10*	*50*	*100*	*500*	*1,000*
1	0.02	0.01				
2	0.04	0.02	0.01			
3	0.06	0.04	0.01	0.01		
4	0.07	0.05	0.01	0.01		
5	0.09	0.06	0.01	0.01		
10	0.15	0.10	0.02	0.01		
50	0.25	0.23	0.10	0.06	0.01	
100	0.25	0.25	0.16	0.10	0.05	0.01

*The numbers of individuals are assumed to be constant and the generations discrete.

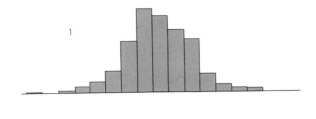

1

2

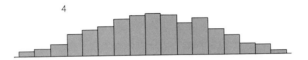

4

8

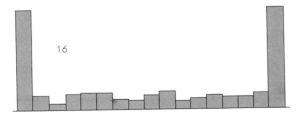

16

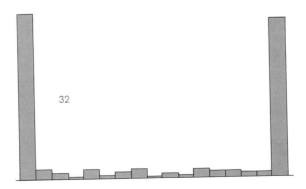

32

Fig. 3.7. Dispersal of gene frequencies among 400 hypothetical populations ("Monte Carlo" simulation on a high-speed digital computer). The following conditions were made for the genetic model: (1) each population consisted of only 8 diploid individuals per generation —4 randomly formed pairs; (2) each individual mated but once and the number of offspring produced by each mating varied—a Poisson distribution with a mean of two; (3) selection and mutation were absent; and (4) each population was started with 2 AA, 4 Aa, and 2 aa individuals, so that there was an initial gene frequency of 0.5 for each allele at a single autosomal locus. Four hundred such populations were simulated over 32 generations. The figure depicts the populations classed according to gene frequencies in generations 1, 2, 4, 8, 16, and 32. Because of chance variation in such small populations, there is an increasing spread in gene frequencies toward fixation of one allele or the other in most of the populations. Data courtesy of Henry Schaffer.

When we average the gene frequencies in a number of similarly endowed subpopulations, we find that the average does not change from one generation to the next but that the variance of the values increases with time. If, however, a single population of the group is considered, the gene frequency q fluctuates from generation to generation until the allele is fixed or lost. This phenomenon of chance alteration due to sampling error is known as genetic drift (the importance of genetic drift in natural populations will be discussed in Chapter 6).

Inbreeding

The outstanding condition for the Hardy-Weinberg model, that there be random mating in large populations, is actually an ideal situation seldom met in reality. Individuals tend to mate with others nearby and, depending upon the extent of the area occupied and the mobility of the individuals concerned, a large population becomes more or less subdivided into smaller interbreeding subpopulations. We have seen in the preceding section that gene frequencies in small populations tend to drift due to gamete sampling error. There is a tendency toward the fixation of one or another allele at any segregating locus. Consequently: (1) there is a reduction in genetic variance within a small population (the individuals become more alike in genotype); (2) there is a general decrease in over-all heterozygosity simply because heterozygotes are in greater proportion at intermediate gene frequencies compared to the more extreme values, which are the ones approached due to random changes; and (3) since the process of drift is random, related subpopulations will become differentiated, depending on which alleles become common or rare in each. We previously presented the idea of random gene-frequency changes in terms of sampling variance. The idea may also be given in terms of inbreeding. It should be apparent that restrictive mating in small subpopulations results in mating pairs having a high degree of coancestry. Although mating may be at random (phenotypically nonassortative), mates will be more closely related to each other in small populations than in large ones. The genetic significance of inbreeding is the consequential increase in homozygosity. Of importance in this respect is the fact that rare recessive alleles will occur in greater homozygous frequency in inbred populations compared to large, random-mating populations. For this reason, plant and animal breeders have made wide use of systematic inbreeding programs. The value of such programs is that they produce homozygous individuals which, when selected, will breed true (see James L. Brewbaker's *Agricultural Genetics,* in this series).

A population subjected to a regular system of inbreeding (repeated selfing, sib mating, etc.) becomes divided into a series of reproductively independent lines (subpopulations). In any one line the increase in homozygosity is accompanied by changes in gene frequencies and a tendency toward the fixation of one or another allele at each locus; the gene frequencies in a series of lines derived from the same base population will tend to diverge. But, when all similarly inbred lines are treated together as a single "population," assuming none are lost or discarded, the effects of inbreeding are independent of the average gene frequency of all lines. In this light, the degree of inbreeding is just an expression of the way the alleles are combined as genotypes— the proportion of homozygotes is greater in inbred groups.

This may be illustrated by following the genotypes of one locus (alleles A and a) over generations in a "population" under a strict system of self-fertilization. Either homozygote, AA or aa, produces only one kind of gamete with respect to the locus in question. The union of two gametes from a single homozygous individual must carry like alleles, ignoring rare cases of mutation. Therefore, the genotype of an offspring of any homozygote is the same as that of its parent. But a heterozygote produces both A gametes and a gametes and hence three classes of offspring, which occur on the average in the ratio $\frac{1}{4}$ AA, $\frac{1}{2}$ Aa, and $\frac{1}{4}$ aa. Because crossbreeding is ruled out, homozygotes cannot contribute to the formation of heterozygotes, and since only half of the progeny of the heterozygous individuals are like themselves, we see that the amount of heterozygosity among the offspring averages half that of the parent generation. There is a corresponding and evenly divided increase of the two homozygous classes. The reduction of heterozygosity is expected to continue over generations of selfing until all genotypes are homozygous. Consider for instance a random-mating population at equilibrium with $p = q = \frac{1}{2}$ that suddenly and completely converts to self-fertilization as the mode of reproduction. The genotypic proportions in the initial population are $\frac{1}{4}AA$, $\frac{1}{2}Aa$, and $\frac{1}{4}aa$. The two homozygous classes contribute half of the progeny in the next generation ($\frac{1}{4}AA$ and $\frac{1}{4}aa$) and the heterozygotes produce half of the progeny, which are distributed in the familiar 1:2:1 ratio, or $\frac{1}{8}AA$, $\frac{1}{4}Aa$, and $\frac{1}{8}aa$. In sum, the genotypic frequencies in the first generation are $\frac{3}{8}AA$, $\frac{1}{4}Aa$, and $\frac{3}{8}aa$. The genotypic proportions of the progeny formed by all the self matings over successive generations are given in Table 3.4. Note that the percentage of heterozygotes is halved each generation.

There is a rapid approach to complete homozygosity upon selfing, which is the most intense form of inbreeding. As pointed out above,

Table 3.4   The decrease of heterozygosity under systematic self-fertilization, starting with an equilibrium population ($p = q = \frac{1}{2}$). The gene frequency does not change with inbreeding when all the selfed progeny are considered together. The inbreeding coefficient, F, increases over generations until all the individuals are homozygous. Since the original gene frequencies are each $\frac{1}{2}$ in this example, it is expected that when F equals 1, the proportions of the two homozygous classes will be equal. Note that the approach to $F = 1$ is progressively slower as inbreeding proceeds over generations.

	Genotypic frequencies				
Generations	A/A	A/a	a/a	F	q
0	$\frac{1}{4}$	$\frac{1}{2}$	$\frac{1}{4}$	0	$\frac{1}{2}$
1	$\frac{3}{8}$	$\frac{1}{4}$	$\frac{3}{8}$	$\frac{1}{2}$	$\frac{1}{2}$
2	$\frac{7}{16}$	$\frac{1}{8}$	$\frac{7}{16}$	$\frac{3}{4}$	$\frac{1}{2}$
3	$\frac{15}{32}$	$\frac{1}{16}$	$\frac{15}{32}$	$\frac{7}{8}$	$\frac{1}{2}$
4	$\frac{31}{64}$	$\frac{1}{32}$	$\frac{31}{64}$	$\frac{15}{16}$	$\frac{1}{2}$
n	$\dfrac{1 - (\frac{1}{2})^n}{2}$	$(\frac{1}{2})^n$	$\dfrac{1 - (\frac{1}{2})^n}{2}$	$1 - (\frac{1}{2})^n$	$\frac{1}{2}$
∞	$\frac{1}{2}$	0	$\frac{1}{2}$	1	$\frac{1}{2}$

Fig. 3.8. The percentage of homozygous offspring from systematic matings with different degrees of inbreeding. After S. Wright, *Genetics,* 6 (1921), 111–78.

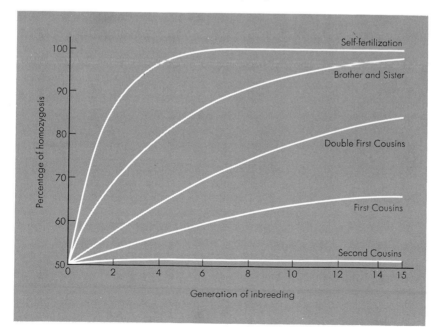

inbreeding may also occur in bisexual forms if the mating partners have a familial relationship (e.g., if they are sibs, half sibs, or cousins). Although matings involving relatives also lead to a reduction of heterozygosity, the loss is much less per generation than under selfing (Fig. 3.8).

The degree of inbreeding in a population is usually measured by Wright's inbreeding coefficient, F, which expresses the amount of heterozygosity that has been lost. The genotypic proportions in populations with different amounts of inbreeding are given in Table 3.5

Table 3.5.   The genotypic frequencies in random-mating populations with no inbreeding, partial inbreeding, and complete fixation

		Genotype frequencies		
Generations of inbreeding	F	A/A	A/a	a/a
None (Hardy-Weinberg)	$F=0$	p^2	$2pq$	q^2
One or more (Wright's equilibrium formula)	$1 > F > 0$	$p^2 + Fpq$	$2pq - 2Fpq$	$q^2 + Fpq$
Infinite number (complete homozygosity)	$F=1$	$p^2 + pq$	0	$q^2 + pq$

(also see the values of F calculated over generations of selfing in Table 3.4). Wright's general equilibrium formula for the genotypic proportions, $AA : Aa : aa$, is $p^2 + Fpq : 2pq - F2pq : q^2 + Fpq$. Note that for a given amount of inbreeding the heterozygotes are reduced by $F2pq$, a quantity half of which is added to each class of homozygotes. In a noninbred population F is zero and Wright's values reduce to the familiar Hardy-Weinberg case.

Besides expressing the average inbreeding of all members of a population in a particular generation, the inbreeding coefficient is a measure of the probable allelic identity of a single inbred individual. By allelic identity we mean that the two alleles representing a locus in a diploid organism are *identical by descent*, i.e., the two gametes forming the zygote come from related parents and carry alleles that are replicate products of a single gene in a common ancestor. Let us clarify this point. Any autosomal locus will be represented twice in a diploid organism. The two genes may exist in the same or in different allelic states, either homozygous or heterozygous. Although the two alleles of a homozygote are considered alike in structure and function ("alike in state"), they may or may not have descended from the same ancestral allele. If they originate from different sources (from similar alleles in unrelated individuals) they are said to be independent, whereas they are *identical* if they are replicates of one and the same allele existing in some past generation. A homozygote may be independent or identical. The value F is the expected proportion of identical homozygotes. It is an expression of the increase of homozygosity over that which would occur if all mates were unrelated and possessed independent alleles.

The concept of allelic identity imparts the precise meaning of inbreeding. We can reconsider the average degree of inbreeding in a population (presented above) in terms of allelic identity by rewriting Wright's proportions in another equivalent way.

$$\begin{array}{ccc} & \textit{Independent} & \textit{Identical} \\ AA & p^2(1-F) + & Fp \\ Aa & 2pq(1-F) & \\ aa & q^2(1-F) + & Fq \end{array}$$

Here we see that a fraction of the population, F, is made up of identical homozygotes and the remainder, $1 - F$, consists of independent homozygotes and heterozygotes. Let us note at this point that all the individuals of any population are related in some degree. Thus, a meaningful measure of inbreeding must relate back in ancestry to some base (individual or population) beyond which allelic identity will not be traced further; the base is a point in lineage at which all alleles are considered essentially independent. In most studies, the base is either a large random mating population or one or more individuals choosen at random therefrom.

The inbreeding coefficient may be calculated under certain conditions for random-mating finite populations. The formulae for ascertaining the degree of inbreeding in a given generation and the loss of heterozygosity due to inbreeding are quite comparable to Eq. 3.5 (see Strickberger, 1968, Chapter 32, for a detailed description). It should suffice here to point out that the amount of inbreeding in populations, like the variance in gene frequencies among subpopulations, may be deduced from N, the effective population size. Although the inbreeding coefficient may be calculated for given generations of a population, it is seldom done in practice. Inbreeding measurements are more commonly determined for individuals and are based upon pedigree analyses.

The value of F for an individual depends on the degree of genetic relationship of his parents and can be measured directly from his pedigree. For example, let us calculate the chance that an offspring is homozygous for identical genes when his parents are sibs and his two grandparents are non-inbred. Here we specify that the grandparents are the base; they are unrelated and carry independent alleles, which may be designated A_1 and A_2 in the grandfather and A_3 and A_4 in the grandmother. The pedigree and the procedure for determining the probability of gene identity is as follows:

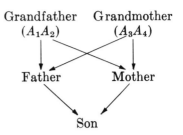

Unrelated Grandparents   Grandfather   Grandmother
                           (A_1A_2) (A_3A_4)

Related Parents (Sibs)   Father   Mother

Inbred Offspring   Son

1. The father receives half of his genes from the grandfather. The chance that gene A_1 is present in the father is therefore $\frac{1}{2}$. If A_1 is present in the father, the chance that it will be passed on to the son is again $\frac{1}{2}$. Thus, the chance that the son will receive gene A_1 from his grandfather by way of his father is $\frac{1}{2} \times \frac{1}{2} = \frac{1}{4}$.
2. Likewise, the chance that gene A_1 is present in the mother is $\frac{1}{2}$, and that it is transmitted from the grandfather to the son through the mother is $\frac{1}{4}$.
3. The chance that the son receives two A_1 genes (i.e., that he is homozygous for identical genes A_1A_1) is the product of the above two probabilities, or $\frac{1}{4} \times \frac{1}{4} = \frac{1}{16}$.
4. Similarly, the chance that the inbred son has two A_2 genes is $\frac{1}{16}$, two A_3 genes is $\frac{1}{16}$, or two A_4 genes is $\frac{1}{16}$.

Since there are four possible ways for allelic identity, A_1A_1, A_2A_2, A_3A_3, and A_4A_4, each with a $\frac{1}{16}$ chance of occurrence, the total probability of allelic identity at a given locus (the value of F for the inbred son) is $4 \times \frac{1}{16} = \frac{1}{4}$. Stated differently, an average of 25 percent of all loci will be homozygous for identical alleles in the progeny of sib parents. The inbreeding coefficient for an offspring from the union of half-sibs (individuals with one common parent) is $\frac{1}{8}$, and that from a cousin mating is $\frac{1}{16}$. You might want to check the validity of these values by constructing representative pedigrees and following the reasoning set forth above (see Victor A. McKusick's *Human Genetics*, p. 103, in this series).

Equilibrium populations and factors of evolution

According to the Hardy-Weinberg model, an "ideal" population of infinite size reaches an equilibrium, or steady state; this is the antithesis of evolution. Organic evolution is a *change* in the hereditary constitution of populations, so that dissimilarities exist between descendant and ancestral forms. We have said that populations evolve, both by progressive changes within lineages and by diversification following separation into multiple distinct units in their descent from a common origin. These changes are quantified into different orders of magnitude. Microevolutionary steps refer to the simple alteration of gene frequencies which occur within a relatively short period of time;

they are reversible and repeatable. Microevolution is so short-termed that it can often be observed directly in nature or in laboratory experimentation. Change resulting in divergence between distinct organic stocks, that is, differences leading to the origin of higher systematic categories such as genera, families, orders, and classes is called macroevolution. These names for the magnitude of modification are descriptive and relative. The mass of phylogenetic and experimental evidence on the causal agencies of evolution supports the theory that all changes are more or less gradual and that highly divergent populations (macroevolution) are formed by the accumulation of individually small steps (microevolution). Based on this conclusion, organic evolution can now be described in more general terms as any progressive change in gene frequencies (the word *progressive* is added here to exclude from the definition reversible changes resulting in minor fluctuations around a stable equilibrium point).

By definition, any factor instrumental in changing gene frequencies is a potential source of evolution. As described above, sampling error during gene transfer from one generation to the next in finite populations disturbs the steady state. Thus, we can classify random genetic drift among the factors which can cause an evolutionary modification of populations. The change δq is nondirected in this case. Other factors can produce directed changes $\overline{\Delta q}$, which are systematic or steady moves toward a particular equilibrium. Directed changes are predictable in both amount and direction, while genetic drift is predictable only in amount. The three factors responsible for directed changes are recurrent mutation, recurrent immigration, and mass selection. These forces can result in a "steady drift." Random processes consist of any fluctuations in the intensities or rates of the above-directed factors, plus sampling error (important only when N is small). These nondirected processes result in "random drift." In summary, the *total change* in gene frequency per generation, denoted by Δq, can be separated into two components: changes caused by random events, and changes resulting from directed processes, where $\overline{\Delta q} + \delta q = \Delta q$.

The above factors are responsible for disturbing the Hardy-Weinberg equilibrium; they were purposely ignored when describing the model. Since few natural populations are free from all these disturbing forces, the existence of equilibrium states seems highly theoretical. But in fact, populations can maintain equilibriums other than that defined by the Hardy-Weinberg law. These other steady states are possible when the forces of one or more evolutionary factors act in opposition. For example, an equilibrium of gene frequencies occurs when mutation tends to increase the frequency of an allele and selection tends to decrease it (see Chapter 7). For any constant set of conditions, the allelic frequencies of all loci will tend toward a genetic equilibrium as, for example, in a singular, unchanging environment; of course, the fixation of one allele at a locus is considered here to be a trivial case of equilibrium.

In sum, two major concepts have been expressed in this extension of Mendelian principles to the level of populations.

1. A state of genetic equilibrium is predicted for large random-mating populations in the absence of forces instrumental to changing gene frequencies.
2. Evolution is defined as a progressive change in gene frequencies. The evolutionary factors known to be responsible for changing gene frequencies are selection, mutation, migration, and genetic drift. These forces work together in such a manner that the population reaches and maintains a moderately steady state as long as the conditions remain fairly constant. There is a tendency toward another equilibrium when conditions change or in some instances when there is chance deviation away from the original equilibrium.

References

Crow, J. F., "Breeding structure of populations. II. Effective population number," in *Statistics and Mathematics in Biology*, O. Kempthorne, T. A. Bancroft, J. W. Gowen, and J. L. Lush, eds. (Ames: Iowa State College Press, 1954), pp. 543-56.

Falconer, D. S., *Introduction to Quantitative Genetics*. New York: The Ronald Press Company, 1960. See Chapters 1 to 6. Contains a brief but one of the most lucid accounts available of the fundamentals of population genetics.

Li, Ching C., *Population Genetics*. Chicago: University of Chicago Press, 1955. An advanced mathematical treatment of genetics at the population level.

Rasmuson, Marianne, *Genetics on the Population Level*. Stockholm, Sweden: Scandinavian University Books, 1961. An excellent basic account of population genetics in a small text.

Strickberger, Monroe W., *Genetics*. New York: The Macmillan Company, 1968. A complete text of basic genetics. See Chapters 30 and 32.

Genetic Variation and Its Source

The significance of biological variation was tacitly expressed in the preceding chapters as follows: evolution results from a change in the hereditary make-up of a species (Chapter 1); the necessity for variation is demonstrated by the overwhelming diversity of life, at the population as well as at the species level (Chapter 2); and no variation results when there is an absence or balance of evolutionary forces (Chapter 3). It is now understood that a multitude of genotypes exists in populations of sexual, outcrossing individuals—variation is considered the rule. Hereditary variation is the product of mutation, the flow of genetic material between populations, and the recombination of genetic factors, all in conjunction with selection pressures which mold the variation pattern. The origin, maintenance, and shaping of variation patterns will constitute the bulk of our remaining discussions. For now, we shall introduce a few general aspects of variation and discuss its source.

Kinds of variation

Variation is just another way of saying that differences exist among individuals. We can speak abstractly of "total" variation, but in practice only one or a few characters is defined. Variation is generally defined according to the population identity of the individuals compared. Codemic members exhibit intrapopulation (individual) variation, whereas interpopulation (geographic) variation refers to differences among individuals representing separate populations. For the latter we simply calculate the average of the individual variation of each group and then compare the averages. For the moment let us concentrate on the variation within demes—individual variation.

When describing variation within populations, we would like to examine and classify all the significant character differences of an unbiased and large sample, but we can never do this because of the vastness and complexity of the variation pattern and the limitations of analytical procedures. The biologist is forced to select those characters, or components of the total pattern, which he deems the most significant for understanding the evolution of the group studied; but, in truth, this choice of characters ultimately depends upon the methods of analysis at his disposal, or more often upon his own personal interests.

Studies of differences among individuals traditionally were based on gross external anatomical features. Over the years, the variation patterns of many populations have been more thoroughly studied, and cytological, histological, serological, physiological, and behavioral characters have been further examined. Recently, biochemical methods (analyses of protein, enzyme, and DNA) have been used successfully to describe previously unknown characteristics. Most comparisons, however, are still based on morphological differences. Morphological variations comprise those structural components that are easily defined, measurable, and classifiable (such as weight, height, and length, color patterns and hues, the number of particular structures, and types of protein structures).

An observed variation of a character is usually either quantitative or polymorphic. The pattern is quantitative when there is a continuous, or graded, series of morphological types, such as those based on size, weight, and skin color in human beings. The variation is polymorphic when two, three, or only a few distinct classes of easily identified "morphs" exist within a population, such as sexual dimorphism, various distinct blood groups, or chromosomal variation. Often it is difficult to distinguish between quantitative and polymorphic characters. In general, the eye color of human beings is referred to as a polymorphism of blue, brown, and black; but an intergrading series of colors occurs to some extent among these classes. Moreover, it is important to realize that the classification of characters as quantitative or polymorphic is somewhat arbitrary, because the genetic basis is essentially the same in principle for both patterns. The difference lies primarily in the number and action of the genetic factors determining the characteristics. Quantitative characters are those controlled by many genes with cumulative effects. Polymorphic characters are more likely to be initiated by the action of "switch" genes, that is, genes whose action causes a switch of the epigenotype to a different developmental pathway. It may also be that a polymorphic character is associated more

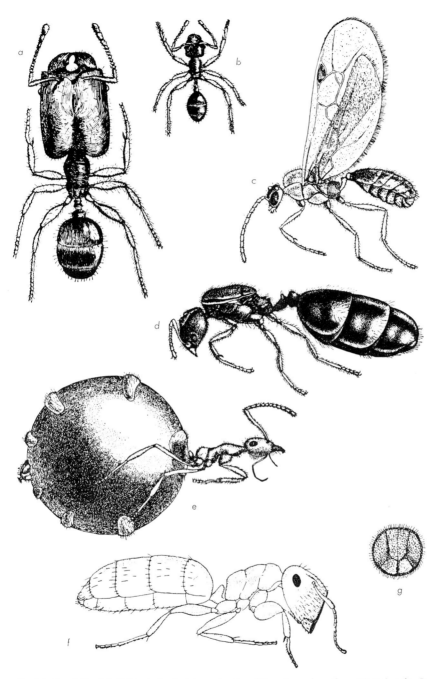

Fig. 4.1. Specialized individuals (castes) in ants. *A*, soldier; *B*, worker; *C*, a winged male; *D*, a female who has lost her wings, of *Pheidole instabilis; E*, a "replete" or honey barrel of *Myrmecocystus hortideorum; F*, a soldier door-keeper of *Colobopsis etiolata;* and *G*, its head in front view. From Theodosius Dobzhansky, *Evolution, Genetics, and Man*, John Wiley & Sons, Inc. (after Wheeler).

directly with the primary product of individual gene action (or of a linked-gene system acting as one functional unit). Both kinds of variation have been amply demonstrated in the majority of the analyses of populations (see James L. Brewbaker's *Agriculture Genetics,* in this series, for a detailed discussion of quantitative inheritance). Polymorphic variation is usually conspicuous, easily classified, and simply analyzed. Thus, the conspicuous illustrations of intrapopulation variation are polymorphisms. The most often quoted examples are the morphologically distinct forms of isopteran and hymenopteran insects (Fig. 4.1).

The adult members of a colony of termites form distinct castes that are structurally specialized for particular functions. The colonies of most species are composed of a primary reproductive caste, a secondary reproductive caste, a worker caste, and a soldier caste. Each group consists of both males and females; however, the reproductive organs of the workers are only little developed and those of the soldiers are not functional. The primary reproductive individuals are sexually perfect winged males and females. They fly from the nest in late spring, shed their wings, and mate. Some couples find a suitable nesting site and found a new colony, of which they are the queen and the king. The abdomen of the queen becomes greatly swollen because of the growth of the reproductive organs and the production of eggs. If either of the royal pair dies, he or she is replaced by one of the members of the wingless secondary reproductive caste, which are sexually mature but nymphal in body form. Almost all labors of the colony are performed by the small, wingless workers, who are blind and sterile. The chief function of the soldiers, who are easily recognized by their enormous heads and mandibles, is the protection of the colony. Some genera do not have the soldier caste. In their place are nasuti, who have small heads characterized by an elongated noselike process that exudes a sticky fluid used for defense and nest building. The immature nymphs of all castes undergo gradual metamorphosis. They are associated in various degrees of development with the adult forms.

Within a single colony of a social species are several classes of morphologically distinct individuals. The termite example was chosen because it illustrates all four basic, or "population specific," kinds of variation related directly to the structure, function, and breeding systems of a population. These are:

1. Age variation (nymphs and adults)
2. Sexual variation (fertile and sterile males and females)
3. Functional, or social, variation (reproductives, workers, and soldiers)
4. Seasonal variation (winged and wingless primary reproductives)

Of course, the termite, as a social insect with a caste system, is a selected example of extreme individual variation. But it can be said that demes of all sexual species exhibit some individual variation. Much of this is of the "general" type, which can occur within a caste, sex, age, and/or seasonal group. By general we mean any variable characteristic (such as blond, red, and brunet hair color in human beings) that cannot be shown to be directly associated with the particular morphological or physiological features needed for the perpetuation of the population. This does not imply that such characters are necessarily nonadaptive, however. It may well be that genes directing the development of these traits also confer different degrees of adaptedness directly or indirectly upon the individual. We simply mean that the characters are not included in the description of the life history of the species.

Genotype–phenotype

It is the significance of variation, not its classification, that is important for understanding the forces of evolution. A brief discussion of Wilhelm Johannsen's terms *genotype* and *phenotype* may help to pinpoint this significance. Character variations within populations are expressed as phenotypic differences. A phenotype is the sum total of morphological, physiological, ecological, and behavioral attributes of an individual during any and all stages of its existence. It is the observed pattern. The phenotypic expression of an organism, continuously changing from the zygote stage until death, results from the interaction of environmental factors with the directing influence of its genotype. The genotype, or sum of the ancestral hereditary material received in the zygote, determines the developmental course of the individual within a particular environment.

The same genotype may direct different developmental pathways and produce different phenotypes under diverse environmental conditions (Fig. 4.2). For example, sexual dimorphism in the majority of animals is genetically determined by the kinds and numbers of chromosomes carried by the individual, but this is not always true. The sex of the echiuroid worm *Bonellia viridis* is determined by the environs of an individual during its larval development. Larvae influenced by a chemical substance on the surface of the proboscis of a mature female develop into males, and free-swimming larvae become females. The morphological dissimilarities of the sexes are spectacular. Males lack digestive organs and are so minute that several can live, in parasitic fashion, within the genital ducts of a single female. On the other hand, the femaleness of the sterile freemartin in cattle is determined by the directing influence of the chromosomes it carries (genetic), whereas its maleness is brought about by intrauterine hormone action initiated by its twin brother (environmental).

The potential phenotypic expression of a given genotype, considered in relation to all environmental situations under which that genotype can survive, is its reaction range, or *norm of reaction*. Certain "norms" are relatively narrow, meaning that if the genotype exists at all, it is predictably constant in phenotypic expression (developmental canalization). On the other hand, when the reaction ranges are wide, development is more flexible and diverse phenotypes are produced under

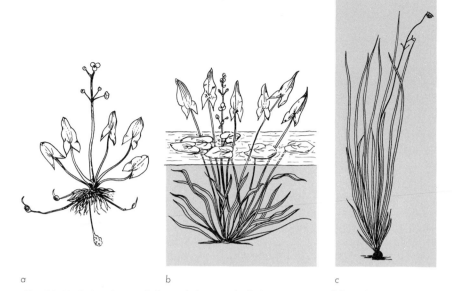

a   b   c

Fig. 4.2. Variations in morphology of the arrowleaf, *Sagittaria sagittifolia,* when it grows on land, when partially submerged, and when completely submerged. These variations result not from hereditary differences but from interaction of the plant's genotype and its environment. The thin leaves developed under water lack a cuticle and absorb nutrients from the water; they have no structural strength and would collapse if not buoyed. In contrast, the terrestrial leaf stands erect; if buffeted by water currents, it would probably be torn off. The root system of land plants is better developed than that of the aquatic form; nutrients are taken from the soil. This variability of arrowleaf undoubtedly has a genetic basis. Many land plants die when submerged, and many aquatic plants cannot cope with terrestrial life. Adaptive responses such as those of arrowleaf may thus be selectively advantageous under certain circumstances. From B. Wallace and A. M. Srb, *Adaptation,* 2nd ed. (Englewood Cliffs, N.J.: Prentice-Hall, Inc., 1964), p. 98.

different environmental conditions. Also, different genotypes can result in similar phenotypes when reaction ranges overlap. The study of developmental processes and their importance to evolutionary theory is just now becoming appreciated. For example, contemporary controversies between botanists and zoologists over speciation patterns in plants compared with those in animals can better be evaluated when it is realized that animals are mobile, seek out their environment (behavior flexibility), and thus tend to be more specialized and less variable than plants. On the other hand, most plants are intimately tied to their environment and must possess enough developmental flexibility (wide "norms") to exist where they locate. At the population level, genetic variability of many plant species, additional insurance for meeting novel environments, is extremely valuable. Thus, the separate evolutionary forces must be evaluated not only in relation to each other but also with reference to the "kind" of species studied and its particular way of adjusting to the environment.

Considering all norms in a population, phenotypic variation (represented by V_P) can be separated theoretically into genetic variation

(V_G) and environmental variation (V_E). Experimentally this division of phenotypic variation requires a knowledge of the genetic relationship (cousins, sibs, parent-offspring, etc.) among individuals. The degree of genetic relationship is reflected by increased phenotypic similarity among related individuals. Under various simplifying assumptions, which must include a consideration of the interactions between the genes and of the interactions of genes with the environment, the amount of increased phenotypic similarity proportionately reflects the genetic variation. Such calculations are very complicated, and their refinement is a continuing problem of quantitative genetics.

Genetic variation involves differences among reaction ranges, whereas environmental variation refers to phenotypic differences within ranges. Nongenetic variation, often called *phenotypic plasticity*, allows the individual to better adapt itself during development. Thus those genotypes with wider reaction ranges are considered better adapted in general because they can develop over a greater range of environmental situations; such genotypes are more versatile. We shall see in Chapter 7 that greater amounts of genetic variation confer similar properties upon the population.

Although variability has been amply demonstrated in most populations, we must also recognize that the norms of all the individuals of a population have much in common (overlap). Canalization, which is the ability of developmental pathways to compensate for genetic and environmental differences, tends to force all the genotypes of a population in the direction of a single phenotype, that is, one specifically adapted for existing conditions. Much of the genetic variation in natural populations is hidden behind a plastic phenotype (often referred to as the wild type)[1] resulting from the similarities of norms and the effects of canalization.

Factors affecting genetic variation

By now it should be clear that understanding the mechanism of evolution as a process requires a basic knowledge of variation in populations, particularly genetic variation. At the time Darwin and other early evolutionists were at their peak in literary production, a major plank in their thesis was that heritable variation is needed for phyletic advance. At the same time they assumed that natural selection tends to produce singularity of forms (the "type" that is particularly adapted). Nearly all hereditary variation was thought to be continuously lost. In addition, the basic idea of inheritance at that

---

[1] We saw in Chapter 2 that emphasis on the similarity of closely related forms usually leads to typological thinking; however, here the interpretation of similarity allows for ample genetic variation and subsumes that deviates are a natural and expected attribute of a varying population.

time was the "blending theory." A little reflection will reveal that with blending inheritance among sexual, cross-fertilizing forms, the genetic variance is reduced by one-half each generation unless there is a high correlation between the hereditary endowment of mates (Fisher, 1930). Knowing that heritable variance would be lost by blending and by selection against the nonadapted forms, the only reasonable inference is that the genetic variation needed for an evolutionary advance is of recent origin and that, whatever its source, the rate of production is immense. In addition, the newly arisen types have to be adapted to existing conditions or they too are immediately lost because of selection —the prime prerequisite for the Buffon-Lamarckian viewpoint, which suggested that new forms are produced as a *direct* response to the environment (the inheritance of acquired characters). These ideas were also responsible for the mutationist's theory that evolution occurs through major steps by way of gross gene or chromosomal changes leading to new functions and structures.

The recent view as to the source, maintenance, and loss of genetic variation is a much modified form of these views, influenced by Mendelian genetics. We have seen in Chapter 3 that with particulate inheritance there is no basic tendency for genetic variance to be lost (Hardy-Weinberg theory). Depletion results only through directed changes in allele frequency by selection or through nondirected changes by genetic drift. Now it is assumed that populations can more easily maintain genetic variance and can generate many genotypes with varying degrees of adaptedness. Those forms produced each generation on the nonadapted end of the scale are weeded out by selection, but they are continuously produced and are considered to be "stores on hand" which would become the progenitors of future generations in the event of a changed environment. They are "preadapted" genotypes ready for new situations which might be met by the population. This should not imply that there is a possible environment for all deviates. Some could not exist under any conceivable set of conditions. However, the group becomes extinct when no genotypes exist which are preadapted for a novel set of conditions under which the population is shifted.

In summary, the population must maintain enough genetic variation (preadapted types) or it will face possible future extinction, but on the other hand the population must not produce an overabundance of deviates in any one generation or it will run the risk of immediate ill-adaptedness. The problem confronting the population is thus the need for *immediate fitness* superimposed upon the need for enough genetic *flexibility* to allow the gene pool to alter when necessary. Some sort of balance between these diametrically opposed necessities must in some way be maintained by the population. Before we see how different forms effect a balance, let us discuss in general the basic factors which determine the amount of genetic variation in populations. First we shall explore the source of genetic variation.

Mutation, the source of original genetic variation

Heredity, unlike evolution, is conservative. Heredity leads to similarities in familial descent, whereas evolution denotes change. Variation within a lineage becomes apparent only when there has been a deviation from total inheritance, that is, when some offspring is not a perfect copy of its progenitors.

It is now commonly believed that the hereditary information ("genetic code") resides in particular base-pair sequences along the chain(s) of the genetic material DNA, which in most higher organisms is organized in the form of nucleoproteins into a still higher functional unit, the chromosome. One of the greatest advances in biology was the discovery of an intrinsic property of the chromosome, now expressed in terms of the DNA molecule—its ability for precise replication during successive cell generations. Herein lies the essence of heredity. Original variation is possible only when the orderliness of the reproduction process falters. Actually, changes take place in the DNA molecular structure, some of which result in new base-pair sequences that can initiate specific metabolic actions different from the action of the original structure. Most often the new molecule is approximately as stable as the original one, and it too replicates in a precise and orderly manner throughout cell generations. Such alterations are referred to as *gene mutations*, and each mutation involves that section of the DNA molecule called the muton.[2] However, any "discontinuous change with a genetic effect" (Mayr, 1963) can be considered a mutation. Although vague, this definition is general enough to cover not only the basic chemical changes at the molecular level but also aberrations at the chromosomal level.

Just as individual loci mutate, the chromosomes themselves are subject to gross alterations, in which whole chromosomes or parts thereof may be lost or multiplied, and spatial arrangements of chromosomal segments may be reorganized. It is convenient to classify the various kinds of chromosomal aberrations as either numerical changes or structural changes.

Structural changes of the chromosomes and the supergene

The amount of genetic variation in a population is dependent on the genetic information present, which is a function of the types

---

[2] The new ideas about the definition and delineation of the "gene," i.e., muton, cistron, and recon (Benzer, 1957; see Philip Hartman and Sigmund R. Suskind, *Gene Action,* in this series), acquired primarily by work with microorganisms, have not been evaluated sufficiently for higher organisms and have not infiltrated population genetics to a point which would warrant their use. The classical terms *genes* (loci) and *alleles,* defined operationally in reference to any study, are still appropriate for discussion at the population level. An excellent review of this topic is given by Grant (1964).

of alleles at all loci, their frequencies among individuals, and their particular combinations as genotypes. There is little doubt that gene mutation, forming the different allelic states, is the source of original variation. But the ways in which the alleles are combined in the chromosomes is also an important source of additional differences. The reason for this must be sought in the way the genetic material is structured (see Franklin W. Stahl, *The Mechanics of Inheritance,* in this series) and the way it functions (see Hartman and Suskind, *Gene Action*).

Classically, the coded genetic information was considered to be an integrated system of discrete genes, serially ordered along the chromosomes and simply defined as those sections of the chromosome separable from others by crossing over. In addition to being defined structurally, each gene was further defined as a functional unit which directed a prescribed development sequence resulting in some discernible aspect of the phenotype—a trait. We have retained much of this concept today. The genes *are* serially ordered along the chromosome, crossover data are still honored, and the genetic sites are still thought to have a primary function (directing the formation of amino acid sequences of enzymes, which in turn specifically guide the chemical reactions in a developmental sequence).

However, the concept breaks down when the functional unit and the structural unit are considered as but one. With the discovery of pseudoallelism in higher forms as well as in microorganisms has come the realization that the functional gene is divisible by rare crossing over. The functionally defined gene is composed of subgenes, which have been revealed in the *cis-trans* test of complementation.

In addition, biochemical studies have shown that genes are combined into higher-ordered gene systems. Separate genes act in a stepwise fashion along a particular sequential biosynthetic pathway, such as in the production of the amino acid arginine in *Neurospora*. Although each gene works independently in a particular reaction step, all the genes of any developmental sequence are thought to work together. They cooperate to elaborate a final product. The functionally related set of genes interact and are said to form a "serial gene system" (Grant, 1964). The genes making up such a system may be scattered throughout the genome or may be contiguously linked over a section of one chromosome. In fact, it has been demonstrated precisely in *Salmonella,* and convincingly in some other forms as well, that in their sequential expressions in biosynthesis certain genes are "clustered." This arrangement is similar to that of the multiple subgenes of a single pseudoallelic locus, since the contiguous subgenes are likewise functionally related. However, complementation is characteristic of the separate members of a linked serial gene system, whereas subgenes of the same cistron have greater functional unity.

However confusing the many definitions of the gene may be, they are also enlightening. Gene interactions and the continuity of the genetic material along the chromosome have gained new importance. The chromosome is no longer thought of as a mere bag of independent loci. The chromosome is made up of genes and subgenes that are serially ordered structurally and are functionally related by different degrees. A hierarchy of functional units exists (from the basic nucleotides, to subgenes, to genes, to serial gene systems, to the total material of the genome). Changes at any of these structural-functional levels culminating in a genetic effect is by definition a *mutation*.

A chromosomal mutation involves the breakage and structural rearrangement of the chromosome. The genes are realigned in new combinations. This may result in a change in phenotype, despite the fact that the loci themselves have not been altered. Any expression such as this that is dependent only on the spatial arrangement of the genes and that often mimics in expression but does not involve gene mutation is termed a *position effect*. Not all reorganizations of the chromosome result in position effects. Many do, but a large percentage of these are usually adaptively detrimental to their carriers.

Chromosomal mutations, or aberrations, are structural rearrangements of the chromosomes. Their production is preceded by breaks of chromosome arms and is completed by the union of the pieces in new combinations. The basic types of chromosomal aberrations are deficiencies, duplications, inversions, and translocations (Fig. 4.3).

Deficiencies and duplications

The haploid chromosomal complement contains one complete set of genes, or one genome, which works as an integrated group in development. The viability or normal functioning of a gamete is possible only if the integrity of the genome is not disrupted. Two such sets of genes normally are basic to the development of the diploid phase, although some aberrations that are detrimental when homozygous can exist in the heterozygous state, somewhat in the same way that the effects of inferior recessive gene mutations can be concealed by their dominant alleles. The loss (deficiency) or gain (duplication) of any euchromatic section of a chromosome alters the dosage of the genes. If the development of an organism is controlled by the direction of coded genetic information present in the DNA portion of the chromosomes, it is easy to see that the destruction or loss of even a small part of that "programmed information" would play havoc in the developmental process. In fact, many deficiencies genetically act as recessive lethals. Duplications, on the other hand, only increase the dosage of certain genes, and their effect on development is quantitative. They are more tolerated in the genome.

These kinds of aberrations are of interest in evolution for two reasons. First, gametes with duplicated and deficient loci may develop after crossing over in chromosomes of individuals that are heterozygous for inversions and translocations (discussed in the following section). This means that these heterozygotes are poorly fertile unless

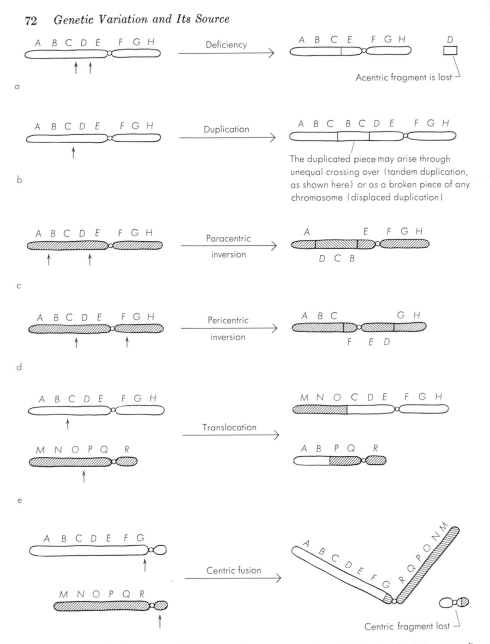

Fig. 4.3. The basic types of chromosomal rearrangements (a–f). The standard (or normal) chromosomes are shown on the left and the structurally altered chromosomes on the right. The small arrows mark the breakage points on the normal chromosome which are required to produce the corresponding rearrangement.

they have compensating mechanisms which aid in preventing development of aberrant gametes. Second, duplications and ploidy (multiplication of whole chromosomes) represent the only known means by which the genetic material can increase. The addition of new loci and the divergence of their functions are reasonable expectations in view of evolutionary progression from the more simple to the more complex forms of life.

Inversions and translocations

If single breaks occur in two separate nonhomologous chromosomes, the distal ends may exchange their positions and become attached to the reciprocal centric units. This type of two-break exchange is a translocation. New linkage groups are obviously created by such a transposition of previously unlinked loci; these become evident once tested genetically. If the exchanged pieces are of unequal length, a translocation is also cytologically detectable during mitosis. Translocations are also revealed by the fact that four chromosomes are associated at pachytene during meiosis.

Whole-arm translocations result from breaks very close to the centromere of each chromosome. A special kind of whole-arm translocation is the centric fusion, which results in the attachment of the major arms of acrocentric chromosomes (also called rod-shaped chromosomes; the centromere is very close to one end). A centric fusion requires a break near the centromere in the large arm of one chromosome and a second break in the minute arm of a nonhomologous one. The whole-arm acentric piece then combines with the whole-arm centric fragment, yielding one large chromosome. The position of the centromere in the new chromosome depends on the relative size of the original acrocentric chromosomes. Two small fragments are also present. Even though one contains a centromere, both are usually lost during division. This happens most often when the fragments possess only heterochromatic material.

Since two acrocentrics "fuse" to form a single chromosome and since a centric fragment is subtracted from the chromosomal complement, the most significant effect of fusions is a reduction in chromosome number. Such a process is known to have occurred in *Drosophila*. Although the primitive haploid number of chromosomes in this genus is six, the basic number of several species is five, or four, or even three. In many species it has been possible to determine the precise chromosomes that have combined.

A chromosome is broken into three pieces when two breaks occur in it simultaneously. Only one of the three pieces contains the centromere. If the middle section turns around and its broken ends fuse with the unrelated ends of the two distal portions, the genes of the internal part will be in reverse order compared with the original sequence. Such a change is called an inversion. If the middle section contains the centromere, the inversion is described as *pericentric,* while an inversion formed by breaks in only one arm of a chromosome is called *paracentric*. In either case the linkage relationship of the genes would be different. In fact, inversions were originally found through linkage studies.

Paracentric inversions do not include the centromere. They can be identified in only a few organisms that have conspicuous pachytene chromosomes by observing the pairing configurations in the heterozygous state. It is also possible to infer their existence when certain types of abnormal disjunction and chromatid bridges occur at meiotic anaphases. But by far the most rewarding studies of paracentric inversions have been of *Drosophila*. Insects of the order Diptera, especially the species in the genus *Drosophila*, are noted for possession of giant polytene chromosomes in the cells of the salivary glands. Not only are the salivary gland chromosomes fantastically large, but they are also aperiodic structures, with distinctive bands, puffs, and other diagnostic "landmarks" along their length (Fig. 4.4). Once these banding patterns are familiar to the experienced cytologist, they can be used to identify any chromosome, every minute section of it, and assuredly any alteration of the banding sequence that results from an aberration. All but the most insignificant rearrangements can be mapped on the giant chromosomes. Cytologists working on *Drosophila* have been additionally blessed by the fact that the homologous chromosomes exist in perfect and permanent synapsis—each band pairs with its homologue. A structural change in a chromosome results in a different sequence of the bands. When such an altered chromosome is paired with a normal one in a heterozygous individual, the chromosomes must bend, loop, and otherwise compensate for their differences so that each band aligns with its homologue. A specific pairing configuration is characteristic for each kind of aberration (Fig. 4.5). Deletions and duplications

Fig. 4.4. A photograph showing the aperiodic nature and synapsis of polytene chromosomes (chromosome III of *Drosophila mojavensis*). Although seemingly one, this is actually two homologous chromosomes intimately paired, band-for-band, along their lengths. Compare with Fig. 4.5, where the separate chromosomes may be detected. Courtesy of Jean C. Brown and L. C. Saylor.

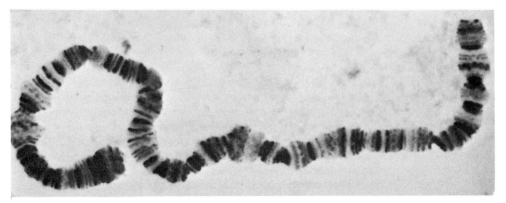

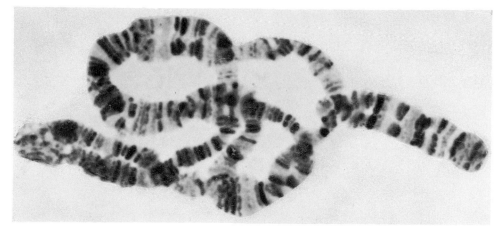

Fig. 4.5. Synapsis of two homologous chromosomes differing by two overlapping inversions. These are chromosomes (chromosome III) of a hybrid individual from the cross of *Drosophila mojavensis* and *D. arizonensis*, each species having its own characteristic structural arrangements of chromosome III. Courtesy of Jean C. Brown and L. C. Saylor.

form buckles, because a portion of one chromosome has no homologous section with which to pair. Heterozygous inversions are characterized by loops, and translocations appear as crosses.

Such complicated configurations likewise exist during synapsis in the germ line. Imagine what happens when crossing over and disjunction take place in the knotted chromatids of the heterozygote. We might guess that the consequences and their effects on recombination and gamete formation are not the same for the different types of aberrations. Let us look at the case of a heterozygous translocation. The synapsis of homologous regions of the four chromosomes (two pairs associated with the translocation) results in a tetrad of all four in the form of a cross (Fig. 4.6). At the first meiotic anaphase, if the chromosomes disjoin and two each go to the separate poles at random, duplication-deficiency segments will then exist in two-thirds of the gametes. Such gametes are nonfunctional. The only balanced genomes are products of the assortment of "parental" combinations, i.e., when alternate chromosomes (chromosomes opposite each other in the cross configuration) go together into the same nucleus.

Cytologically heterozygous chromosomes can also differ in their particular combination of alleles. The parental gene combinations are recombined by crossing over in translocation heterozygotes just as they are in any cytologically homozygous individual, except that the genes near the original breakage points are more tightly linked.

This is not true for an inversion heterozygote. When a chromosome with a "standard" arrangement of genes, $A B C D E F G H I$, is present in an individual having another chromosome with a paracentric inver-

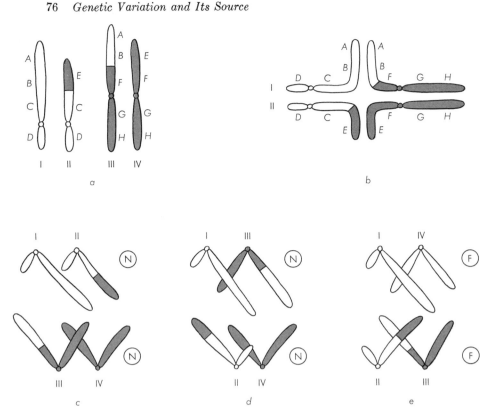

Fig. 4.6. Pairing and disjunction of chromosomes in a translocation heterozygote. (*a*) The chromosome constitution of the two homologous pairs of chromosomes involved in the translocation. (*b*) These chromosomes are expected to synapse as indicated (chromatids are not shown). The group of four chromosomes may disjoin in one of three ways (*c, d, e*) at anaphase during the first meiotic division. If adjacent chromosomes go to the same pole (I and II, and III and IV, as in *c;* or I and III, and II and IV, and in *d*), the resulting gametes will possess duplication and deficiency segments, and will be nonfunctional. If alternate chromosomes go together to the same pole (I and IV, and II and III, as in *e*), then each gamete will receive a complete genome and will be functional. One can demonstrate this by labeling the anaphase chromosomes according to the genetic regions shown in *a* and *b*.

sion, $A\ B\ C\ D\ \overline{G\ F\ E}\ H\ I$, a characteristic loop configuration is expected at meiosis. Crossovers outside the region of the loop produce recombination of any differing allele combinations. The chromosomes separate normally, and each becomes incorporated into separate daughter nuclei. However, a single exchange within the confines of the inversion leads to the connection of the two centromeres by a chromatid bridge, which stretches and usually breaks at anaphase. The other exchange product is an acentric fragment (Fig. 4.7), which is eliminated from the division spindle. Duplication-deficiency gametes are then produced (but note that their formation is initiated by crossing over with inversions, whereas nondisjunction is the cause with translocations). Crossing over also produces partial sterility of pericentric inversion heterozygotes.

Translocation or inversion heterozygotes are expected to have a reduced fertility due to the production of duplication-deficiency gametes. But both inversion and translocation polymorphisms exist in natural populations. Certain races of the plant genus *Oenothera*, for example, are composed solely of translocation heterozygotes. These forms have developed a mechanism to insure the alternate disjunction of chromosomes during meiosis, and therefore have circumvented the cytological difficulties associated with translocations. Inversion heterozygotes are even more common in nature. They have been found extensively in the genus *Drosophila* where detection is facilitated through salivary gland chromosome analysis. The fertility of inversion heterozygotes is not impaired in this genus because crossing over does not

Fig. 4.7. The three possible genotypes in a population segregating for a standard chromosome and a paracentrically inverted sequence are shown in *a*, *b*, and *c*. Synapsis and crossing over are quite regular in the case of either the standard-chromosome homozygote or the inversion-chromosome homozygote. A loop formation (*d*) is expected in the synapsis of the heterozygote, compensating for the structural difference between the homologues. Crossing over, involving two of the four chromatids, is indicated between regions C and D. Such a crossover will lead to the formation of a chromatid bridge and an acentric fragment during Anaphase I (*e*). The fragment is usually lost in the cytoplasm, while the bridge may or may not break. Only those nuclei of the second division that receive an intact standard chromosome, *A B C D E F*, or the complete inverted chromosome, *A E D C B F*, are expected to give rise to functional gametes. Gametes receiving the crossover products will possess duplications and deficiencies and will not be functional.

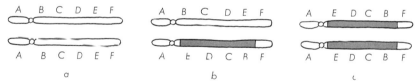

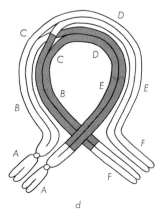

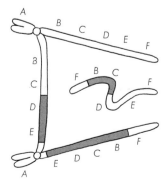

take place at all in males, and oogenesis is organized in such a manner that most crossover chromatids are not included in the functional nucleus of the egg. But why should a species develop such cytological complications in the first place? What is the advantage of an inversion? The linkage of adaptive allele combinations appears to be the answer.

The supergene

Because recombinational gametes are nonfunctional and because crossing over is reduced in the inverted segments of heterozygotes, all the genes included therein are inherited as one completely linked group. Although the genes on the inverted segment are not necessarily functionally related, as those of a linked serial gene system are, they are adaptively codependent and are selected for or against as a group; the group is a single physical and functional unit in evolution. Such a linked set of loci can be thought of as a supergene, or single giant locus. Alleles in the inverted region do not recombine with those of the noninverted sequence of a homologous "locus." The different arrangements may be considered "allelic," and their relative abundance in a population can be expressed as allele frequencies.

Recombination incessantly generates a phenomenal array of allele combinations in a cross-fertilizing species, certain of which direct the formation of very desirable phenotypes; but recombination can also destroy these harmonious groups. Structural mutations are also continuously produced, but only those that happen to tie together a group of adaptively superior genes will be retained in the population, though this does not occur very often. Once it has, however, the new "allele" will be reproduced and accumulated. In fact, many populations are known to be polymorphic for two or more superalleles. *Drosophila willistoni* has the greatest number; more than 50 inversion "morphs" have already been discovered.

If the superiority is expressed in the homozygous state, the inverted arrangement may even replace the "standard" sequence, as demonstrated by the comparison of many related species (as well as races and populations), which, though structurally homozygous, prove to differ from one another because of rearranged banding patterns. In other words, different supergenes have become fixed in them. It may be that some of these differences arose and the inverted superallele replaced the normal allele after the species diverged. Another model suggests that structurally homozygous species can evolve from a single polymorphic ancestral population. The species-specific arrangements are derived from those originating in the polymorphic population, and fixation is assumed to take place in isolates which later diverge as separate species.

To reiterate, gene mutation is the ultimate source of most variation. Chromosomal mutation, although possibly associated with a position effect, is mainly a mechanical device whose effect is associated with recombination and linkage.

The fate of a single mutant

Gene mutations and aberrations differ in another important way. The chemical change involved in the transition from one allelic state to another recurs time and again. Each kind of change is characterized by a specific mutation rate. On the other hand, any chromosomal rearrangement is considered an extremely rare, probably a unique, event. We have seen that salivary gland chromosomes have a very detailed gross morphology. Over **1,000** distinct bands have been detected in the X chromosome of *Drosophila melanogaster*. Extrapolating to the full chromosomal complement, it is estimated that about **5,000** microscopically differentiable regions exist. Since most spontaneous rearrangements result from two breaks that can come about nearly at random along the chromosomes, the chances are slight that a rearrangement will ever be repeated by breaks at the same precise points. Chromosomal mutations do not have a measurable mutation rate—each may be considered to be a single occurrence.

What is the fate of a single mutant? Fisher (1930) suggested the following model and conclusions: let us imagine a large finite population composed exclusively of AA homozygotes and assume that a novel mutation to a occurs, resulting in a single heterozygous (Aa) individual. Will the mutant gene increase in future generations? The fate of the new mutant gene will depend on the number of offspring that the heterozygote engenders and on their worth in the population. At this stage it is convenient to let the "worth" be inconsequential, i.e., selection is inoperative. That leaves us with family size as the major determinant.

It is obvious that the mutant gene is immediately lost from the population if the heterozygote leaves no progeny. Even when it contributes to the next generation, only half of all progeny fostered are expected to carry the mutant allele (the heterozygote mates necessarily with an AA homozygote). This means that for each family size (n), the probability of extinction of a is $(\frac{1}{2})^n$.

Family size (n)	0	1	2	3	$\ldots$	n
Probability that a is not present (L_n)	1	$\frac{1}{2}$	$\frac{1}{4}$	$\frac{1}{8}$	$\ldots$	$(\frac{1}{2})^n$

The exact family size produced by the heterozygote is of course not predictable. However, we know that the various classes (0, 1, 2, 3, $\ldots$, n) should have a mean of two when the population size is not changing. Fisher pointed out that for certain family-size data the frequency of the various classes corresponds rather closely to a Poisson distribution (the mean and the variance is approximately **2**).

Family size (n)	0	1	2	3	$\ldots$	n
Frequency (F_n)	e^{-2}	$2e^{-2}$	$\dfrac{2^2}{2!}e^{-2}$	$\dfrac{2^3}{3!}e^{-2}$	$\ldots$	$\dfrac{2^n}{n!}e^{-2}$

By multiplying the chance of extinction (L_n) by the frequency of occurrence of the corresponding family class (F_n) and summing all the probabilities, it is unimaginable but mathematically correct that the total probability of the loss of a in one generation is e^{-1}, or 0.3679. The probability that the mutant will still exist is then $1 - 0.3679$, or 0.6321. For example, if 100 separate unique mutations occurred, nearly $2/3$ of them would survive in the progeny generation. The process recurs the following generation but starts with only 63 mutants. Of the 100 only 53 are expected to remain after the second generation of chance removal. Eventually all are lost. Even those mutants with a slight selective advantage are subject to extinction by this chance process. About the only way a novel aberration could become established in a large population is by having a significant selective advantage at its inception (i.e., the "worth" we ignored above).

Gene mutations arise many times in the history of the population. The mutation rate acts as a force (or "pressure") which continues to supply identical mutant genes to the gene pool, even when they are detrimental.

Dynamics of gene frequencies under recurrent mutation pressure

As we studied in Chapter 3, the evolutionary status of a population from a genetic viewpoint is characterized by the relative frequencies of alleles (gene frequencies) at individual loci in the population at a given time. Mutation as an evolutionary force directs the change of gene frequency by shifting one allelic state to another at a prescribed rate. The consequence of such changes in gene frequencies is in turn reflected in the pattern and amount of genetic variability of the trait controlled by these genes. Some changes in genetic variability are cryptic, as is demonstrated by recessive mutant alleles whose frequency is so low that they exist almost exclusively in heterozygotes. This potential variation (stored) will become free variation (expressed) as recurrent mutations accumulate and mutation heterozygotes segregate. Free variation is manifested by the phenotypes and becomes the raw material of natural selection.

Although mutation and natural selection cannot be divorced in the actual evolutionary process, it will be less complicated at the outset to eliminate selection when considering the simple quantitative relationship between the mutation rate and the gene-frequency changes. The mutation rate is usually measured by the probability for one allele to mutate per generation. In other words, the mutation rate is 1/100,000 (or 10^{-5}) when 1 in 100,000 alleles of one kind mutates in one generation. Mutation rates are functions of the kinds of alleles and of both external and genic environmental factors. It is often found that the rate is about 10^{-5} or 10^{-6} in natural environments. However, much higher rates have been reported (for example, 10^{-4} for neurofibromatosis in man and for certain isozymes in Drosophila). As we might suppose from such a low value of mutation rates, the significance of recurrent mutations in the evolutionary process is far greater when the population size is large (see Chapter 6). For this reason, we shall again consider only theoretical populations of infinite size.

One-way mutation (irreversible mutation)

Let p_0 be the gene frequency of allele A in the initial population and u the mutation rate for the change from A to another allele a. Let p_1 stand for the frequency of A after one generation. According to the definition of mutation rate, the gene frequency of A will decrease by the amount $\Delta p = up_0$. Thus we have

$$p_1 = p_0 - \Delta p = p_0 - up_0 = p_0 (1 - u) \qquad \text{(Eq. 4.1)}$$

Advancing one more generation, the gene frequency p_2 is obtained by p_2 and p_1 in the places of p_1 and p_0 in Eq. 4.1. Thus

$$p_2 = p_1 (1 - u)$$

and since we know $p_1 = p_0 (1 - u)$,

$$p_2 = p_0 (1 - u)^2$$

Repeating this process t times, the gene frequency after t generations will be

$$p_t = p_0 (1 - u)^t \qquad \text{(Eq. 4.2)}$$

This means that allele A would eventually disappear from the population. Since u is very small in comparison with 1, $(1 - u)^t$ can be approximated by e^{-ut} (e is the base of natural logarithm 2.718 . . .). Thus, we can write Eq. 4.2 as

$$p_t = p_0 e^{-ut} \qquad \text{(Eq. 4.3)}$$

and

$$q_t = 1 - p_0 e^{-ut} \qquad \text{(Eq. 4.4)}$$

where q is the gene frequency of a. The reader may recognize the great similarity of the last two equations to the simple catalytic equation used in chemistry and to the sigmoid growth-curve equation. This is not coincidental, for in all these equations the frequency of mutant alleles, the concentration of chemicals, or the growth in plant height, all represented by q, start from the beginning value given as relative fraction $q_0 = 1 - p_0$ of the final limit value at time 0 and proceed exponentially toward the final limit 1 as times goes on. The rate of conversion from one state to another is constant (u) for each allele, molecule, or cell division.

In order to visualize the speed of evolutionary progress purely under mutation pressure, let us compute the number of generations required to reduce the gene frequency of A by a factor of $\frac{1}{2}$. The answer is

$t = 0.69/u$ (by substituting p^t in Eq. 4.3 by $\frac{1}{2} p_0$ and taking the natural logarithm of both sides). Thus, with $u = 10^{-5}$, 69,000 generations are required for the half-life of A. Let us consider the case of $p_0 = 0.96$. It will take 69,000 generations to reach $p = 0.48$, another 69,000 generations to reach $p = 0.24$, and so forth. Therefore, such a mutational process alone takes approximately a third of a million generations ($5 \times 69,000$) to change $p = 0.96$ to $p = 0.03$. The vastness of such a great number of generations required for a substitution of one allele for another points out that mutations alone will probably not account for the actual speed of evolutionary changes observed in nature.

Two-way mutation (reversible mutations)

When allele A mutates to allele a at the rate of u and a mutates back to A at the rate of v, the gene frequency p_t is reduced by $- p_t u$ and increased by $(1 - p_t)v$. Thus, the net change Δp_t is equal to $- p_t u + (1 - p_t)v$. The gene frequency of the following generation p_{t+1} is

$$p_{t+1} = p_t - \Delta p_t = v + (1 - u - v)p_t \qquad \text{(Eq. 4.5)}$$

The major difference between the two-way and the one-way mutation processes is that opposite mutations form a system of counterforces. Just as they do in the balanced system between gravity and a coiled spring with a fixed weight, the opposing mutations come to a point of equilibrium. At the point of equilibrium the gene frequency will not change any further ($\Delta p_t = 0$). Thus, equating both p_t and p_{t+1} in *Eq. 4.5* to the equilibrium gene frequency $\hat{p}$ it is found that

$$\hat{p} = \frac{v}{u + v} \qquad \text{(Eq. 4.6)}$$

For example, two-way mutation system with $u = 2 \times 10^{-5}$ and $v = 10^{-5}$ ($A \xrightarrow{u} a$ is double that of $a \xrightarrow{v} A$) will tend toward the equilibrium point of $\hat{p} = \frac{1}{3}$.

Corresponding to Eq. 4.3, the gene frequency after t generations is

$$p_t = \hat{p} + (p_0 - \hat{p})e^{-(u + v)t} \qquad \text{(Eq. 4.7)}$$

where p_0 again symbolizes the gene frequency of allele A at the initial generation. As t increases, $e^{-(u + v)t}$ decreases to zero from the positive direction. When p_0 is smaller than $\hat{p}$, the difference between the initial gene frequency and the equilibrium value is negative. Thus, Eq. 4.7 indicates that p_t *increases* monotonically to $\hat{p}$ when $p_0 < \hat{p}$. Conversely, p_t *decreases* monotonically to $\hat{p}$ when $p_0 > \hat{p}$.

The mathematical theory tells us that mutation is an evolutionary force which can change gene frequencies. The action may result in fixation (allele replacement) or in a balanced equilibrium when operating in a large population not influenced by selection. However, mutation rates are small and mutation pressure is considered a relatively weak force (compared with selection or drift) for changing gene frequencies. In conclusion, the major role of mutation in evolution is the production of variation rather than the shaping of the variation pattern.

Gene flow

New alleles may be added to a gene pool by mutation, as we have seen, or by way of foreign gametes furnished by immigrants from other populations. We call the latter process gene exchange or, if it is a recurrent process, gene flow. Mutation and gene flow are similar processes because both furnish novel alleles or additional alleles to a population—each is a source of variation.

Demes, geographic isolates, races, and even species are seldom closed systems. A certain amount of gene transfer among them is always a possibility, which becomes increasingly probable the closer the engaging populations are related spatially and genetically. Gene flow is highest among the adjacent demes of one species. Since the amount of migration (interbreeding) is high, we can reasonably suspect that contiguous demes possess similar gene pools. Assuming this to be true, it follows that gene flow is not instrumental in altering gene frequencies or in contributing per se to added variation, since the allele frequencies in the migrants would not differ significantly from those in the recipient population.

Geographic isolates (populations of a species separated by geographic barriers) are more independent, gene flow among them is restricted, and selection can lead to local adaptation. Dissimilar alleles become fixed or are present in much different frequencies among the separate populations. Gene flow is by definition a rare event among isolates, but when genes are exchanged, it is a highly significant event —adaptive gene complexes are broken up, allele frequencies are altered, and in general, all genetic differences between the populations are reduced. This is especially true when complete secondary contact is made and gene flow is extensive.

The effectiveness of gene exchange depends on population structures or more specifically, on the amount of migration (migration rate equals M) and the genetic divergence of the participating populations (the magnitude of the difference of gene frequencies).

To view this in more quantitative terms, let us conjure up one more theoretical model. To keep the problem simple, we shall say that effective migration is unidirectional, e.g., migration from a large mainland population to a smaller, semi-isolated, island population. Again all assumptions used in the Hardy-Weinberg model apply, except that migration (gene flow) is occurring at a constant rate M from the mainland to the island. The gene frequency of the a allele is represented by Q on the mainland and by q on the island. The value of Q is considered to be constant (the fraction lost each generation, M, is a representative

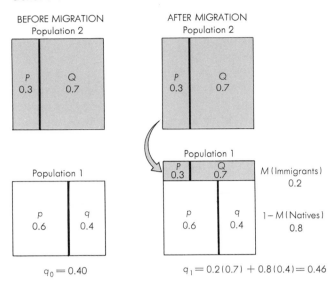

Fig. 4.8. Diagram showing the effects of migration on gene frequencies. It is assumed that the gene frequencies in population 1 and population 2 are different and that migration occurs from population 2 into population 1. After migration, population 1 will consist of 1 — M natives having gene frequencies like those of population 1 before migration occurred, and M migrants having gene frequencies characteristic of population 2.

sample and is extremely small in relation to the size of the mainland population). However, the value of q will change each generation as a greater portion of the island gene pool is represented by migrants and their descendants.

In any one generation a fraction of the island population are migrants M and the remaining individuals are endemic $(1 - M)$. The value of q in each generation is simply the average of the gene frequencies Q and q weighted by the relative fractions of the island population that they represent (Fig. 4.8). Thus

The frequency of a *in generation* $0 = q_0$
The frequency of a in generation $1 = q_1$

$$q_1 = q_0 (1 - M) + QM$$
$$= q_0 - q_0M + QM$$
$$= q_0 - M (q_0 - Q) \qquad \text{(Eq. 4.8)}$$

The change in gene frequency Δq in one generation is $q_1 - q_0$. Substituting the value of q_1 from Eq. 4.8,

$$\Delta q = q_1 - q_0$$
$$= q_0 - M (q_0 - Q) - q_0$$
$$= -M (q_0 - Q) \qquad \text{(Eq. 4.9)}$$

Therefore as stated earlier, the effect of migration, as measured by gene-frequency change Δq, depends on its rate M and the genetic difference of the populations $q - Q$.

The role of gene flow in the development of variation patterns has long been a neglected aspect of both natural and experimental population studies. In fact, most of the latter are deliberately set up as closed systems with special care to avoid contamination. The conclusions from such studies are only indirectly applicable to nature, which is characterized by open populations. However, in recent years the importance of migration and gene exchange has gained new emphasis, primarily through Mayr's (1963) excellent reevaluation of the theory of speciation by geographic isolation (see Chapter 8). The crux of this thesis is that gene flow is a retarding element of speciation. This is certainly true when M is large—divergence among populations is minimized, races combine by interbreeding, and at most, clines are established.

But the migration of foreign individuals into a population may also be considered a source of added genetic variation. This concept is applicable to cases of migration between populations in which the difference in gene frequencies $(q - Q)$ is large, as expected among well-differentiated races or distinct species. For example, Camin and Ehrlich (1958) found that the populations of water snakes, *Natrix sipedon,* on islands in western Lake Erie are variable with respect to the color patterns of the individuals. Virtually all of the *sipedon* population on the Ohio and Ontario mainlands are banded, whereas the partially isolated island populations consist of both banded and unbanded forms, the latter being significantly predominant. The investigators showed that the proportion of banded individuals was greater among juveniles in litters produced by captured females than in the adult populations of the islands. It is known that a given pattern does not change in ontogeny; therefore, it was concluded that there must be differential survival in favor of the unbanded types. The islands are without inland waters and the *sipedon* populations are confined to the lake shores, where the snakes are found resting on flat limestone rocks. Here the unbanded individuals are more inconspicuous than the banded ones (by human judgment). It is speculated therefore that selection may occur through the differential action of predators; gulls and herons have been observed eating the young forms. The question arises why selection has not resulted in an unbanded island race. The investigators suggest that this tendency is thwarted by immigration. Banded individuals migrate continuously from the mainland and contribute to the insular populations, although the environment to them is apparently suboptimal. Thus, it may be said that the variation pattern of the island populations originates and is maintained primarily by recurrent migration.

Most cases of hybridization among related species do not lead to gene exchange, because any hybrids produced are seldom adaptive. Despite this well-known fact poorly fit hybrids are occasionally fertile enough to produce some F_2 and backcross progeny. When the original species are reproductively defined, we likewise expect few recombinant progeny to be adaptive. But under certain circumstances a small amount of backcrossing to one or both parental forms takes place over generations. As backcrossing progresses, more and more of the recombinant genotypes become similar to those characteristic of the parental population(s) to which backcrossing has been recurrent. The closer the genotypes approximate those of the "pure" species, the greater are the chances that they will be adaptive and contribute to the gene pool of the "recurrent parent." A few genes (or linked blocks of genes) can be transferred from one species to another by this process of introgressive hybridization. Here we are speaking of mere "gene trickle" among related species. New genes and gene combinations are added to the genetic architecture of the introgressed form by this process, but the relative integrity of the species is supposedly not destroyed (see the example involving species of *Aquilegia* in Chapter 8).

Introgression requires that some hybrids be originally produced. This seldom occurs between well-differentiated species, especially among animals. However, there are numerous examples of closely related plant species that are capable of producing fertile hybrids, at least experimentally. If such cross-fertile forms coexist in nature, it is often found that they are ecologically isolated by being adapted to specific conditions in neighboring habitats—conditions unfavorable for the existence of hybrids. Although hybrids may be produced, they are seldom instrumental in initiating effective gene exchange between the species. However, the potential is there, and gene exchange can occur if the environment becomes modified in any way suitable for the survival of the hybrids and their derivatives.

The question of how significant introgression is in evolution has not been settled. So many examples of introgression have been reported in the higher plants that it must be considered more than insignificant. Nevertheless, it can be argued that a small number of these examples are really hybrid "swarm" formations between geographical (ecological) races, which are lackadaisically accepted as morphospecies. Since related species of higher animals make choices, insemination mistakes among them are rare. In addition, the hybrids that might be occasionally produced are usually sterile and weak. This is just another way of saying that animal morphospecies are more apt to correspond to "good" biological species than morphologically defined plant forms are. Nevertheless, a few cases of introgression have been found (see the discussion of the towhee in Chapter 2), and no doubt many others have gone undetected.

Recombination as a source of variation

We have been discussing the origin of gene and chromosome variation that is brought about ultimately by mutation and is redistributed among populations by gene flow and introgression. But the variation exposed to the action of selection is expressed by phenotypes, which in turn are reflections of genotypes. The immediate source of variation at the genotype level is not mutation but the reassortment (recombination) of the mutant alleles that have previously accumulated and are stored in and among populations. Parental gene combinations in the gametes are combined at fertilization. These combinations are broken up during meiosis when the nonhomologous chromosomes independently assort (Mendel's second law) and when crossing over occurs between linked loci.

Through the mixing of gene combinations by recombination, an enormous number of genotypes can be generated from a small number of differing alleles among but few loci. If the number of segregating loci is n and the number of alleles at each locus is r, the number of genotypes that can be engendered by recombination is

$$\left[\frac{r\ (r+1)}{2} \right]^n$$

As an example, if each of only 10 loci are represented by 4 possible alleles, the theoretical number of diploid genotypes that can be produced is 10 billion! The formation of many genotypes can come about either by many segregating loci (n) or by many alleles at each of fewer loci (r), or both.

To better grasp the importance of recombination as a source of variation, let us reflect on an example in which it is lacking, namely, in an asexual organism. All variability in asexual forms depends on mutation. The formation of a particular favorable gene combination must await completion until all component alleles have been produced by successive mutations within a single lineage, since mutations in separate clones cannot be interchanged without sexual reproduction. Here we see the significance of sex; it is a device by which gene exchange is possible among separate lines of descent; mutations can be pooled and then shuffled into a multitude of genotypes by recombination.

Asexual microorganisms are able to adapt to slowly changing conditions because they reproduce rapidly and build up enormous populations, thereby compensating for low mutation rates. A few preadapted types are produced each generation by mutation alone. In higher organisms smaller population sizes, longer life span, and developmental complexities, together with evolutionary flexibility, are possible only with sexual reproduction as defined by recombination.

Recombination appears to be a basic property of DNA, and it is not surprising to find some form of it in most organisms (for example, in sexual or parasexual reproduction). Most of the existing asexual higher forms are believed to have been derived from sexual ancestors (Stebbins, 1960). Many of these are not obligately asexual; those that apparently are have left the mainstream of evolutionary advance because of their lack of flexibility.

References

Camin, J. H., and P. R. Ehrlich, "Natural Selection in Water Snakes (*Natrix sipedon* L.) on Islands in Lake Erie," *Evolution, 12* (1958), 504–11.

Carlson, Elof Axel, *The Gene: A Critical History.* Philadelphia: W. B. Saunders Company, 1966. Many of the basic concepts dealing with mutation are given in historical sequence.

Ronald R. Fisher, *The Genetical Theory of Natural Selection.* London: Oxford University Press (Clarendon), 1930.

Grant, Verne, *The Architecture of the Germplasm.* New York: John Wiley & Sons, 1964. Necessary reading for all students of genetics.

Mayr, Ernst, *Animal Species and Evolution.* Cambridge, Mass.: Harvard University Press, 1963. See Chapter 7 for numerous examples of the various kinds of variation in natural populations.

Pontecorvo, G., *Trends in Genetic Analysis.* New York: Columbia University Press, 1958. A good review of parasexuality and recombination in microorganisms.

Stebbins, G. Ledyard, "The Role of Hybridization in Evolution," *Proc. Am. Phil. Soc., 103* (1959), 231–51.

Stebbins, G. Ledyard, "The Comparative Evolution of Genetic Systems," in *Evolution After Darwin,* Sol Tax, ed., Vol. I. Chicago: University of Chicago Press, 1960, pp. 197–226.

Five

Natural Selection

Genetic variation results from mutation and gene flow. During reproduction an array of genotypes develops after recombination of alleles. Of the innumerable genotypes possible, only a relative few are actually formed in any generation during the life of the population. Since all populations are finite and the number of possible genotypes is unimaginably large, the great majority of the theoretical combinations are not realized simply because of chance—even some of those that are produced are later randomly eliminated accidentally. Of course, the smaller the population, the fewer the number of genotypes produced. We saw in Chapter 3 that when populations are very small, alleles may become fixed by the unoriented process of genetic drift. This is one way by which variation is lost from the population.

Although the number of genotypes is also limited in larger populations, there is less probability that the frequencies of alleles will change by chance alone (the Hardy-Weinberg model). When there is competition among individuals, however, some are eliminated differentially rather than accidentally. This discriminative elimination, or selection against certain genotypes, can result in a directed change of gene frequency. Both genetic drift and selection tend to erode variation, but—as we shall find—selection is nonrandom and creates order by increasing the adaptedness of the population; it sculptures new variation patterns.

Let us now examine in more detail (here and in Chapter 6) these factors which reduce genetic variation and reemphasize, just as Darwin postulated, that natural selection is still considered the guiding force of evolution.

Natural selection

The individuals comprising a species differ in many ways. Those that possess certain features which prove useful for their survival and reproduction are considered better adapted. They are naturally selected over the others if there is competition for a limited supply of some factor necessary for life, such as food, a mate, or a place to live. These beneficial characteristics may be associated with the strength of the individual, which would help to insure escape or victory in combat, or they may be more subtle, such as those that increase resistance to disease. They may be even something as seemingly insignificant as the added aerodynamic perfection of the airfoil of the wing of a maple seed. The important aspect of an adaptive trait is that in some way, no matter how slight, it increases the chance that the possessor will contribute offspring to the effective population of the next generation. Of course, survival is an important aspect of adaptedness, because a form must survive to the reproductive stage in order to leave any offspring, but it is only one of many factors determining reproductive success. For instance, altruistic behavior (having regard for the well-being of others) may become an adaptive feature if the action is directed toward relatives and results in a greater number of surviving offspring. This is so even if the altruistic display decreases the probability of survival of the altruist. For example, consider the preferential feeding of young by parents that may themselves be starving in times of food scarcity, or a case in which a parent lures a predator away from the den wherein hides its young. The "survival of the fittest" is not always a selfish matter.

The features conferring greater adaptedness will, if inherited, accumulate over generations, and the hereditary constitution of the population will gradually become altered. The ensuing change is phyletic evolution, the goal is the maximization of the total fitness of the population, and the force responsible is natural selection. Natural selection is simply the nonrandom differential perpetuation of varying genotypes, the results of which can be viewed in two ways: (1) the positive creation of a new or stabilized variation pattern by the differential perpetuation of some genotypes and (2) the loss of variation by the preferential elimination of other genotypes. Of course, these are mutual aspects of but one process.

All populations, including those of man, domestic animals, field crops, and laboratory stocks, are subject to natural selection. In the Darwinian sense the varying trait is fitness. However, in most domestic populations the choice of the individuals which are to become the breeding stock is determined by the breeder or experimenter. This is artificial selection, the purpose and goal of which are set by man. Fitness in artificial selection is defined in the same manner as it is in natural selection, but the concept is complicated by the fact that Darwinian fitness is superimposed upon a commercial value. The distinction is an important one; fitness is of only secondary interest in artificial selection (see James L. Brewbaker's *Agricultural Genetics* in this series).

A variety of levels of reproductive units are found in nature, i.e., genes, gene complexes, chromosomes, individuals, clones, populations, and species. In its classical role (Darwinian) selection occurs among individuals of a population. However, it is also associated with any competition among the reproducible units at other levels of complexity. The genes may have been the units of competition if they existed as separate naked entities during primitive evolution, but this is conjecture. The only known forms which even approximate this state are the subcellular nucleoproteins, the viruses. In cellular organisms the selection of genes or gene complexes is accomplished indirectly by way of the phenotypic expression of whole genotypes. Selection occurs among phenotypes. Inasmuch as genotypic variation is expressed, selection secondarily differentiates among genotypes. In this same sense, it can be said that alleles which help to enhance adaptedness are also selected for. In fact, we usually express the force of selection in terms of allele superiority and gene-frequency changes, although the actual competition is among individuals.

A laboratory demonstration of natural selection

Natural selection is shown in the experiment described below by following gene frequency changes over time in a closed population of *Drosophila*. The demonstration was set up to meet all the criteria of the Hardy-Weinberg model with the exception that natural selection was allowed to operate. Although the population was maintained under controlled conditions in the laboratory, the progenitors of each generation were determined naturally—not by artificial selection.

Two kinds of *Drosophila mulleri* were introduced into a plastic population cage (Fig. 5.1). One stock possessed normal chromosomes, while the other was homozygous for a mutant autosomal paracentric inversion (originally produced by X-irradiation). An inversion was chosen as the phenotypic trait to follow in order to eliminate recurrent mutation from consideration and because each chromosome combination (genotype) was identifiable.

The initial population consisted of 2,000 flies, of which 15 percent were homozygous for the normal chromosome arrangement and 85 per cent were homozygous for the inversion. The population was maintained by introducing fresh media into the cage periodically. The size of the population soon became stabilized at approximately 5,000 to

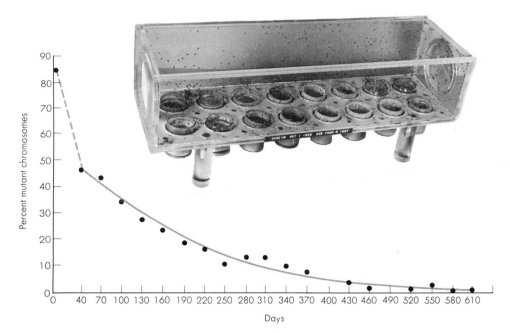

Fig. 5.1. The results of natural selection in a closed population of *Drosophila mulleri*, measured by the change in frequency of a mutant inversion compared with the normal chromosome arrangement (see text). The smooth curve fitted through the observed frequencies represents the expected change in chromosome frequencies, assuming that the relative fitnesses of the normal homozygote, the heterozygote, and the inversion homozygote are 1.0, 0.74, and 0.63, respectively. The photograph is of a typical *Drosophila* population cage. The fly population is maintained over generations by substituting fresh media for old through the exchange of food cups held in rubber gaskets in the bottom of the chamber. The cage is made of Plexiglass and has screened vents at either end. Samples are taken by removing eggs from a fresh food cup and culturing them in vials. Larvae developing therein are dissected for cytological preparations.

6,000 individuals, since food and space were limited. Samples of eggs were withdrawn at 30-day intervals, and the number of homokaryotypes and heterokaryotypes (genotypes) was determined by analyzing the salivary gland chromosomes of the developing larvae. The relative proportions of the normal and mutant chromosomes were then estimated from the genotype frequencies. The percentage of mutant chromosomes for each sample is plotted in Fig. 5.1. The outcome is obvious. In approximately 20 generations (610 days) the frequency of the mutant dropped from 0.85 to 0.0075.

The population was very large (little genetic drift) and closed (no immigration). Mutation pressure was nil. In addition, it was known from previous tests that the three genotypes, normal/normal, normal/mutant, and mutant/mutant, mated nearly at random. Thus the change in frequency is logically attributed to natural selection, since this was the only condition of the Hardy-Weinberg model that was not specified. The reduction of the mutant form is explained tautologically by saying that the individuals carrying the normal arrangement express superior fitness.

Adaptive values and selection coefficients

No two genotypes (individuals) are expected to be identical in sexually reproducing populations. Much of this genetic variation is revealed as phenotypic variation, which is exposed to the action of natural selection. Selection is an expression of the differences of fitness among individuals, which we have said is based on their relative capacities to donate genes to future generations. Now a particular *fitness value* (or adaptive value) ideally should express immediate fitness, or the actual number of progeny produced by each genotype, as well as long-range fitness, or their genetic donation to subsequent gene pools. Although we are now speaking of whole genotypes, the relative adaptive values of genes, gene blocks, or even populations can be expressed with respect to their adaptedness to existing conditions and to their preadaptedness to future situations. This complements the idea that immediate-fitness values relate to the environment in which they have been measured, since an individual's survival and reproductive capacity is correlated with the physical environment (remember the description of reaction norm). Fitness values are strictly relative in still another way. A single value is rather meaningless unless compared with the reproductive performance of the other codemic members. Thus, when discussing selection, we must think of relative fitnesses rather than simply adaptedness.

Describing adaptive values and selection at the level of whole genotypes is getting close to the true nature of the selective process, but it is extremely theoretical and it ramifies the sophisticated concepts of pleiotrophy, polygeny, developmental canalization, homeostasis, and allele and gene interactions in general. Although this approach does characterize much of the modern thinking on selection, speciation, and the adaptive norm (Mayr, 1963), quantification is almost impossible, because each case is essentially unique. Most of our present ideas were built on simpler individual-gene models involving the marginal genotype (defined as the genetic endowment of one *or more* individuals or gametes at a locus under consideration) and changes of allele frequencies at a single locus. Only the immediate aspect of fitness is considered, since there is no practical way of handling the "long-range" concept.

Several individuals may be alike in some portion of their genotypes; they possess a specific combination of certain alleles. Let us say that the locus A segregates for the alternate alleles a_1 and a_2. The population is subdivided into three groups (genotypes) a_1a_1, a_1a_2, and a_2a_2. The adaptive values of the individuals in each group are averaged to give the net fitness values W_0, W_1, and W_2, of the respective genotypes.

Since fitness is relative, one of these values (usually the most favored) can be set at 1, and then the others are expressed as decimal fractions of that standard.

Genotype	a_1a_1	a_1a_2	a_2a_2
Fitness value	W_0	W_1	W_2
or	$\dfrac{W_0}{W_0} = 1$	$\dfrac{W_1}{W_0} = 1 - s_1$	$\dfrac{W_2}{W_0} = 1 - s_2$

The coefficients of selection (s) denote the intensity of selection against the more inferior forms. Each value (s_1 and s_2) stands for the proportionate reduction in procreation of that genotype compared with the standard (a_1a_1 was arbitrarily chosen here). The fitness scale ranges from 1 ($s = 0$), representing the genotype that contributes the most to future generations, to 0 ($s = 1$), in the case of a genotype that is lethal or effectively sterile.

As pointed out in Chapter 3, selection among genotypes may result in a change in gene frequencies. If one allele, let us say a_1, increases in frequency, then it may be said that allele a_1 is selected for and the alternate allele, a_2, is selected against, although the actual process of selection is occurring at the level of individuals, or genotypes. The selection of one sort of allele over another is seldom an all-or-none affair. Some of the individuals possessing a_1 may result in well-adapted phenotypes, while others may be less fit, and this is true also of those with a_2. The reason is that the fitness of any individual would not be determined by these alleles alone: *the adaptive value of each genotype is dependent on the action and interaction of all the genes of the individual.* We see that selection is really a statistical phenomenon with respect to genotypes and the alternate alleles. If, on the whole, the net selective value of all the genotypes carrying a_1 is greater than the value of those possessing a_2, then the a_1 allele will be selected for and will have a higher frequency among the selected individuals compared with the unselected population of each selection cycle. As long as selection favors the a_1 allele, it will continue to increase in frequency each generation. Eventually it should become fixed.

Most higher organisms go through a haplophase and a diplophase in their life cycles. We can imagine one set of fitness values for the gametic stage and another for the diploid genotypes. Haploid fitnesses are affected by such factors as gamete viability or competitive activity in fertilization, while selection may operate through differential viability, relative fertility, or mating propensity among diploid genotypes. Let us first consider the consequences of selection at the gametic stage.

Gametic selection

A random mating population composed of 0.16 a_1a_1, 0.48 a_1a_2, and 0.36 a_2a_2 individuals is expected to produce a gamete pool consisting of 0.4 a_1 and 0.6 a_2 gametes (see Chapter 3), which will unite to form the zygotes of the next generation. However, it is possible that selection may intervene between the formation of the gametes and the formation of the zygotes; the different classes of gametes may differ in their abilities to survive or to otherwise affect fertilization. For example, let us say that the gametes carrying the a_2 allele are only two-thirds as viable, on the average, as those with the a_1 allele. This means that the proportion of a_1 and a_2 gametes that actually form zygotes are equal, i.e., after selection the frequency of a_1 gametes is 0.4 and the frequency of a_2 gametes is $\frac{2}{3} \times 0.6 = 0.4$ or, adjusting the gamete pool to unity, $a_1 = 0.4/0.8 = 0.5$ and $a_2 = 0.4/0.8 = 0.5$. There has been a change in gene frequency from 0.6 to 0.5 for a_2 and from 0.4 to 0.5 for a_1. Such a change will be reflected among the genotypes of the offspring generation, which will become 0.25 a_1a_1, 0.50 a_1a_2, and 0.25 a_2a_2.

Gametic selection may be described in general as follows. A population consisting of p^2 a_1a_1, $2pq$ a_1a_2, and q^2 a_2a_2 individuals will form a gamete pool of p a_1 gametes and q a_2 gametes. If the relative fitness of the superior gamete class, say a_1, is taken as unity, then that of the inferior forms, a_2, is $1 - s$, where s is the selective coefficient, or the amount of selection against them. Multiplying the frequency of each gamete class times its relative fitness we find that the effective gamete pool (after selection) becomes p a_1 and $q(1 - s)$ a_2, which totaled equals $p + q(1 - s) = 1 - sq$. The relative proportions of the a_1 and a_2 gametes after selection are $p/(1 - sq)$ and $q(1 - s)/(1 - sq)$, respectively. It is important to note here that the sum of the gametic frequencies after selection does not equal unity, but rather, $1 - sq$. We divided by this total in order to make the sum of the relative frequencies given above equal one.

The change in gene frequency brought about by selection is usually denoted by Δq, where Δq equals the difference between q_1 (the frequency of a_2 alleles after selection) and q_0 (the frequency of a_2 alleles before selection). Therefore,

$$\Delta q = q_1 - q_0 = \frac{q(1 - s)}{1 - sq} - q = \frac{- sq(1 - q)}{1 - sq}$$

Here we see that Δq is negative. That is, the frequency of the a_2 allele is reduced by gametic selection. If such selection continues over generations, the a_2 allele will be reduced further and ultimately will be lost from the population.

In the higher organisms, gametic selection probably occurs more commonly in plants than in animals. This is because alleles are more likely to be expressed in the more differentiated haploid gametophytes of plants. However, it is not unknown in animals; gametic ratio distortion has been found in *Drosophila* and mice. Selection in the haploid phase is most important in microorganisms and other lower forms, such as algae and fungi, where the haploid phase constitutes a larger part of the life cycle of the organism. On the other hand, selection in higher plants and animals occurs chiefly in the diploid stage—during the development of the zygote into the mature individual.

The general selection model

Let us now follow the change of gene frequencies by assigning different adaptive values to genotypes and assume that selection occurs only at the genotypic level. Although the adaptive value of each genotype includes a consideration of mating propensity, it is easier to visualize the situation if we assume random mating and think of the action of selection as differential survival. This should not detract from the fact that the equations do take into account all selective processes from one generation to the next during all stages of the life cycle.

The consequence of one cycle of selection can be summarized as shown in the table. Note that the proportionate contribution is given

Genotype	a_1a_1	a_1a_2	a_2a_2	*Total*
Frequency in generation 0, before selection	p^2	$2pq$	q^2	1
Fitness	W_0	W_1	W_2	—
Proportionate contribution	p^2W_0	$2pqW_1$	q^2W_2	$\bar{W}$
Frequency in generation 1, after selection	$\dfrac{p^2W_0}{\bar{W}}$	$\dfrac{2pqW_1}{\bar{W}}$	$\dfrac{q^2W_2}{\bar{W}}$	1

by multiplying the frequency of each genotype by its adaptive value. Since selection is occurring, only one (or two) of the adaptive values can equal 1. Since at least one W is less than 1, the total contribution cannot be unity in a mathematical sense. We have designated this total by the quantity $\bar{W}$, which is called the mean fitness of the selected group. It is simply the sum of the proportions of the genotypes after selection; thus,

$$\bar{W} = p^2W_0 + 2pgW_1 + q^2W_2 \qquad \text{(Eq. 5.1)}$$

The amount $\bar{W}$ can also be given as 1 minus the quantity of each of the more inferior types that were eliminated (compared with the standard genotype). This does not necessarily mean that the size of the population is reduced because of selection; the selectively eliminated individuals may well be included in the class of those forms that are normally lost due to overproduction. The increase or decrease in the size of the population depends on many factors, of which the intensity of selection is only one. It does mean that the reproductive capacity of the population is less than it would be if every genotype produced as much as the average individual making up the superior class. This is an important point to remember for our comments on interpopulational selection in Chapter **7**.

As you may recall from Chapter **3**, Eqs. 3.1 and **3.2**, the gene frequency of either allele in generation 1, p' and q', can be ascertained by adding half of the frequency of the heterozygotes a_1a_2 to the percentage of the homozygous class carrying the allele in question. Thus the frequency of the a_2 allele in the selected population is

$$q' = \frac{q^2 W_2 + pq W_1}{\bar{W}} = \frac{q(q W_2 + p W_1)}{\bar{W}} \qquad \text{(Eq. 5.2)}$$

The change in gene frequency during one cycle of selection is then $\Delta q = q' - q$, or by substituting the value of q' from Eq. **5.2**

$$\Delta q = \frac{q(q W_2 + p W_1)}{\bar{W}} - q$$

$$= \frac{q(q W_2 + p W_1) - q\bar{W}}{\bar{W}}$$

and W from Eq. **5.1**

$$\Delta q = \frac{q(q W_2 + p W_1) - q(p^2 W_0 + 2pq W_1 + q^2 W_2)}{\bar{W}}$$

$$= \frac{pq[q(W_2 - W_1) + p(W_1 - W_0)]}{\bar{W}} \qquad \text{(Eq. 5.3)}$$

This expression has three fundamental parts

Part 1 $\boxed{pq}$ Part 3 $\boxed{[q(W_2 - W_1) + p(W_1 - W_0)]}$

Part 2 $\boxed{\bar{W}}$

and each part has noteworthy implications.

In part 1 the change Δq is proportional to the quantity pq provided that the other parts are constant. We find that pq is maximal when $q = \frac{1}{2}$ and minimal when either $q = 0$ or $q = 1$ (genetic selection does not occur in the absence of genetic variation).

In part 2 the change Δq is inversely proportional to $\bar{W}$, provided that the other parts are constant. This means that the higher the mean fitness, the slower the change, and vice versa, for a given set of W_0, W_1, and W_2. In other words the rate of change in gene frequency slows down as the mean fitness increases toward its maximum, and the rate is faster when the population is far away from its "potential" reproductive capability. This is akin to the statement made previously that selection tends to maximize the fitness of the population, i.e., the gene frequency continues to change as long as $\bar{W}$ is not the largest value possible.

In part 3 this quantity, $[q(W_2 - W_1) + p(W_1 - W_0)]$, is the most significant of the three parts. It is called the additive, or "average," effect of gene substitution. It consists of two sections with comparable differences $(W_2 - W_1)$ and $(W_1 - W_0)$, which represent the effects on fitness when one allele is substituted; i.e., $a_2a_2 \to a_1a_2 \to a_1a_1$. The additive effect is an average of these two differences, weighted by q and p. Here we are expressing the fitness value of genes averaged over the genotypes in which they are found. Any change in gene frequency is proportional to the "average" effect. If the substitution of inferior alleles (alleles that on the average result in genotypes with lower adaptive values) by more favored alleles results in increasing the mean fitness of the population, then the additive effect is a positive value. When Δq is zero, the additive effect is likewise zero. The most obvious examples of this would occur when the superior allele is fixed or when W_2, W_1, and W_0 are equal; no selection occurs and gene frequencies remain constant. However, because the "average" effect is also a function of the gene frequencies p and q, it may assume a value of zero, although the adaptive value differences still exist among the genotypes (see the example of heterozygote superiority below).

The W values in the denominator and the numerator of Eq. 5.3 occur as linear functions. This makes it possible to divide all W values by the largest one to obtain fitness values described in terms of selection coefficients, as expressed in the previous section. The value of Δq may be obtained by using either notation of adaptive values.

The change in gene frequencies of multiple alleles at a single locus can be expressed in a similar manner. The change in the frequency of any one allele in an m-allele system is

$$\Delta q_i = \frac{q_i(\bar{W}_i - \bar{W})}{\bar{W}} \qquad \text{(Eq. 5.4)}$$

where q_i is the frequency of the ith allele among $a_1, a_2, \ldots, a_i, \ldots, a_m$. The mean fitness of all genotypes is $\bar{W}$, while $\bar{W}_i$ is the mean of all the values of genotypes which contain the ith allele (i.e., of $a_1a_i, a_2a_i, \ldots, a_ia_i, \ldots, a_ma_i$).

Since fitness is an aspect of the phenotype, the genes we may wish to follow are expected to express some degree of dominance with respect to the fitness character. Dominance on the fitness scale may be *lacking*, *partial*, *complete*, or *overdominant*. In addition, selection can favor either allele, and the intensity of selection can be *partial* (detrimental genotypes) or *complete* (lethal or sterile genotypes). Once we specify the dominance relationship and give some indication of the selection intensity, we can further simplify the general formula for a specific case. To do this, the W's in the formula are replaced by the "selection-coefficient" notation of fitnesses, and p is replaced by $1 - q$. Only a few of the combinations of the conditions mentioned will be given special attention below.

Complete dominance

If selection is against a detrimental recessive homozygote, s is greater than 0 but less than 1; and $W_0 = 1$, $W_1 = 1$, and $W_2 = 1 - s$ (Fig. 5.2a). The change in the frequency of the recessive allele becomes

$$\Delta q = \frac{(1-q)\,q\,[q(1-s-1) + (1-q)\,(1-1)]}{1-sq^2} = \frac{-sq^2(1-q)}{1-sq^2}$$

$$\text{(Eq. 5.5)}$$

Fig. 5.2. Diagram showing the effects of selection with different degrees of dominance with respect to fitness (a–d). The relative fitnesses of the three genotypes are given on the fitness scale in column B. The amount of selection against given genotypes is pictorialized in column C (relate the large square to those shown in Chapter 3), and is shown in algebraic form in column D. The population fitness, W, is unity before selection and is decreased by the proportion of the genotypes selected against after selection. The expected change in gene frequency for each model is given under column E.

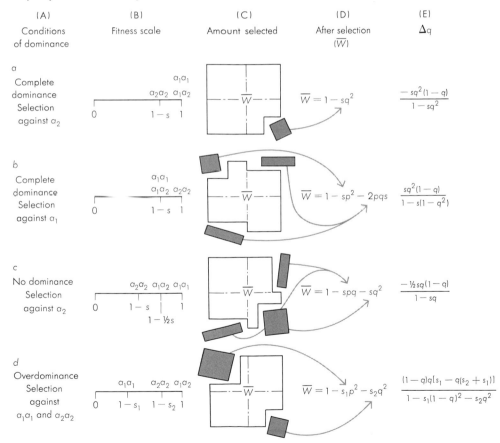

Therefore Δq is always negative, and the frequency of the recessive allele a_2 decreases until it is lost.

When selection is complete, the homozygotes carrying the recessive allele are lethal or effectively sterile; the value of s is 1. Equation 5.5 then simplifies to

$$\Delta q = \frac{-q^2(1-q)}{1-q^2} = \frac{-q^2(1-q)}{(1+q)(1-q)} = \frac{-q^2}{(1+q)} \quad \text{(Eq. 5.6)}$$

If the selection coefficient s or the gene frequency q is very small, the denominators of Eqs. 5.5 and 5.6 (as well as those of other selection models) is nearly unity, i.e., W can be approximated by 1. Then the numerators alone can be used to express Δq, which in Eq. 5.6 is simply $-q^2$. This form is of interest since the reduction in frequency of a rare recessive lethal gene is equal to the frequency of lethal zygotes formed. Therefore the rate of elimination becomes exceedingly slow as q becomes small. For example, the number of generations required to decrease q from 0.01 to various smaller frequencies are

Frequency	0.01000	0.00500	0.00250	0.00125
Number of generations	0	100	300	700

In general it can be shown that the frequency is halved in the course of $1/q$ generations. The process is even slower when selection is only partial. You may judge the relevance of a negative eugenics program involving compulsory sterilization ($s = 1$) of individuals with undesirable recessive defects!

Selection is much more effective against dominants (Fig. 5.2b) than against recessives with the same gene frequency and selection intensity, since the heterozygotes are also exposed to the process of elimination ($W_0 = W_1 = 1 - s$, and $W_2 = 1$.) The gene-frequency change is

$$\Delta q = \frac{(1-q)q\{q[1-(1-s)] + (1-q)[(1-s)-(1-s)]\}}{1-2(1-q)qs-(1-q)^2s}$$

$$= \frac{sq^2(1-q)}{1-s(1-q^2)} \quad \text{(Eq. 5.7)}$$

which for a dominant lethal becomes

$$\Delta q = \frac{q^2(1-q)}{1-1+q^2} = \frac{q^2(1-q)}{q^2} = 1-q \quad \text{(Eq. 5.8)}$$

This expression shows that when only the recessive form survives and reproduces, the gene frequency goes from q to 1 (the a_2 allele is fixed) in a single generation.

Overdominance

When a locus exhibits complete, partial, or no dominance, selection will continue to act until the more inferior allele is eliminated from the population; the sign of the formulae of Δq will either be $+$ or $-$, contingent on whether the a_2 allele is favored or selected against. However, when a locus shows overdominance (the heterozygote is the superior genotype), both alleles are expected to remain in the population and the gene frequency will tend toward an equilibrium at some value between 0 and 1, which is determined by the relative magnitudes of the inferiority of the two homozygotes. We can express the relative fitnesses of the genotypes of an overdominant locus as $W_0 = 1 - s_1$, $W_1 = 1$, and $W_2 = 1 - s_2$ (Fig. 5.2d); and from Eq. 5.3, the change in gene frequency is found to be

$$\Delta q = \frac{(1-q)q\{q[(1-s_2)-1] + (1-q)[1-(1-s_1)]\}}{1 - s_1(1-q)^2 - s_2 q^2}$$

$$= \frac{(1-q)q(-s_2 q + s_1 - s_1 q)}{1 - s_1(1-q)^2 - s_2 q^2}$$

$$= \frac{(1-q)q[s_1 - q(s_2 + s_1)]}{1 - s_1(1-q)^2 - s_2 q^2} \qquad \text{(Eq. 5.9)}$$

The sign of the quantity of Δq is the same as the sign of the additive effect, $s_1 - q(s_2 + s_1)$. When the gene frequency reaches the equilibrium value, denoted by $\hat{q}$, there will be no further change and Δq is zero. The equilibrium value $\hat{q}$ can be ascertained by setting the additive effect at 0 and solving for q

$$0 = s_1 - q(s_2 + s_1)$$

$$q = \frac{s_1}{s_2 + s_1} = \hat{q} \qquad \text{(Eq. 5.10)}$$

The sign and the degree of Δq is a function of the difference between q and $\hat{q}$. If $\hat{q}$ is below the equilibrium frequency, Δq is positive and the gene frequency increases toward $s_1/(s_2 + s_1)$; when q is greater than $\hat{q}$, it decreases until the gene frequency reaches the equilibrium point. This is presented for particular cases in Fig. 5.3. A stable equilibrium results when q reaches $\hat{q}$. An equilibrium is considered *stable* when a gene frequency deviating from $\hat{q}$ in either direction tends to move back toward the point of $\hat{q}$, as in Fig. 5.3, if selection favors the heterozygous genotype. A state of equilibrium is *unstable* when a gene frequency deviating from the equilibrium value in at least one direction tends to move away from the point of $\hat{q}$. Selection which does not favor heterozygotes results in unstable equilibrium (s_1 and s_2 in Eq. 5.9 would then be negative). You may wish to verify this by formulating the change in gene frequency when W_1 is less than both W_0 and W_2.

The gene-frequency equilibrium associated with heterozygote superiority means that each kind of allele can be retained in the population at high frequencies. This may be true even if one homozygote, say

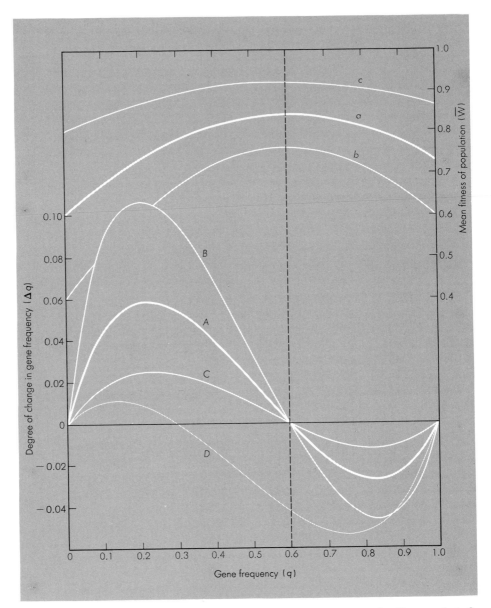

Fig. 5.3. The overdominance model ($W_0 = 1-s_1$; $W_1 = 1$; $W_2 = 1-s_2$), with curves A, B, C, and D showing the relation of Δq over gene frequency. The graphs are based on Eq. 5.8 with the following respective values of s_1 and s_2: (A) 0.4, 0.267; (B) 0.6, 0.4; (C) 0.2, 0.133; (D) 0.15, 0.35. Curves a, b, and c show corresponding $\overline{W}$ values for the first three. The equilibrium value ($\hat{q} = 0.6$) is the same for A, B, and C, since the ratio s_1/s_2 is similar (Eq. 5.9); example D has a lower ratio and a lower value ($\hat{q} = 0.3$). In all cases Δq is positive when q is less than $\hat{q}$, and negative when q is greater than $\hat{q}$. Note that $\overline{W}$ is maximum at equilibrium. Also, the approach to equilibrium is more rapid the greater the selection intensity (compare A, B, and C). Calculations courtesy of Henry Schaffer.

a_2a_2, is extremely inferior, because the favored heterozygote would continuously supply the a_2 allele pool by segregation. The retention of both alleles by overdominance is one method by which polymorphism is maintained at the phenotypic level (see Chapter 7). When the major genes or supergenes and their allelomorphs are selectively balanced, the attainment, or approximation, of an equilibrium state creates a *balanced polymorphism;* the most extreme example is the *balanced-lethal system,* in which s_1 and s_2 are equal to 1 and only the heterozygotes contribute progeny. (What would be the gene frequencies in a balanced-lethal stock? Your answer should help to emphasize that $\hat{q}$ is a function only of the selection coefficients.)

The determination of adaptive values

The mathematical model presented in the preceding section was based on qualifying assumptions, which included some suggestions of the relative adaptive values of the genotypes. Predictions of gene-frequency changes were made with regard to different degrees of dominance for the fitness character. Such information helps us to interpret what occurs in true populations. Nevertheless, the approach in studies of living populations is quite the reverse of that in model building, since it is the selection coefficients that are unknown and since the collected data are more apt to express the quantity Δq. An estimate of the selection coefficients and the corresponding adaptive values are calculated from the observed differences of the genotypic proportions before and after the selection process takes place. This can be ascertained more precisely for individual components of fitness than for total fitness values.

Components of fitness are the qualities of the individual of a certain genotype that influence in any way its final adaptive value (characters such as viability, fertility, fecundity, or sexual activity). Each component is a fitness "trait." Standardized techniques exist for measuring each of these attributes for those forms that can be mated and maintained in the laboratory or experimental garden. A determination of the relative viabilities of the chromosomal types of *D. mulleri* presented earlier in this chapter may serve as an example.

The three karyotypes (or chromosomal types, or genotypes) are designated here as $++$, $+i$, and ii, where i denotes the mutant inversion sequence and $+$ represents the normal arrangement. The viabilities measured between the egg and the adult stages have been determined by observing the differential survival of a known collection of the three chromosomal types under competitive conditions. A large number of heterozygous individuals were produced from the mating of the two stock cultures and were introduced into a population cage containing fresh food. Since only one kind of genotype was present, the relative mating behavior and fecundity were of no consequence. Since all the flies were heterozygotes, it could be assumed that 50 percent of the eggs deposited on the food surface were heterozygous $+i$, 25 percent were homozygous ++, and **25** percent were homozygous ii.[1] Adult individuals emerging in the cage in the next generation were isolated and crossed separately to known homozygotes. The genotype of each fly was then determined through progeny testing by analyzing the salivary gland chromosomes of several larvae that each fostered in the individual test cultures. It was found that the genotypes in the progeny generation (after selection) had frequencies of **0.35** ++, **0.48** +i, and **0.17** ii. It follows that differential survival (selection) occurred among the genotypes during the egg-to-adult period of development.

Now that we know the proportions of the genotypes both before and after selection, we can assign relative survival values to the three forms as shown in the table. Dominance for the survival character is ap-

Genotype	++	+i	ii
Frequency before selection	0.25	0.50	0.25
Frequency after selection	0.35	0.48	0.17
Relative survival values	$\dfrac{0.35}{0.25} = 1.4$	$\dfrac{0.48}{0.50} = 0.96$	$\dfrac{0.17}{0.25} = 0.68$
or	$\dfrac{1.4}{1.4} = 1.0$	$\dfrac{0.96}{1.4} = 0.7$	$\dfrac{0.68}{1.4} = 0.4$
Selection coefficients	0	0.3	0.6

parently lacking because the inferiority of the ii homozygote (0.6) is twice that of the heterozygote (0.3).

Although viability is a major aspect of fitness, it is not equivalent to the adaptive value of a genotype. Theoretically an adaptive value represents the combined measures of all the characters that influence fitness. A few studies have been made in which several of the conceivable adaptive features have been separately measured for a given set of genotypes, but no successful method has yet been devised for combining the individual quantities as total selective values. Also, such "component" studies are often carried out under conditions dissimilar to the environment of the population, and thus the values obtained in each test are not necessarily comparable to the same feature in the population itself.

The only known way in which complete adaptive values can be defined is from data expressing a series of gene-frequency changes over several generations. This kind of information is presented for the *D. mulleri* experiment in Fig. 5.1. Each point in the graph represents the frequency of the mutant arrangement at a given time during the study. To establish each point, the population was sampled, the proportions of

[1] These proportions were actually obtained among well-developed larvae that were raised under optimum conditions.

the individuals of the different chromosomal types were determined, and the gene (arrangement) frequencies were calculated from the zygotic percentages. This is a standard procedure in studies of both natural and experimental populations.

It is a common misconception, however, that the relative adaptive values of the genotypes can be derived from the data of only one of the samples by comparing the *observed* genotypic frequencies with those *expected* by the Hardy-Weinberg Model. By expected we mean the set of genotypic frequencies constructed on the basis of the binomial expansion p^2, $2pq$, q^2 by substituting the gene frequencies *determined from the single sample*. We carried out this procedure in Chapter 3 for the M-N blood types of a group of Australian aborigines and found that the observed distribution of genotypes did not deviate significantly from that expected from the calculated values of p and q. However, Wallace (1958) points out that if a pertinent difference were to exist (although it is true that the Hardy-Weinberg equilibrium does not hold), we cannot use the differences (observed versus expected) as a basis for comparing the adaptive values of the genotypes in question. The reason is that there are two variables involved, the gene frequencies before selection and the fitnesses of the genotypes, neither of which is known. If we use the method to gather information about adaptive values, then we must assume that the "gametic frequencies have remained constant from the beginning of the generation sampled." This is an unwarranted assumption unless it is known (through *repeated* sampling) that there is an equilibrium among the alleles at the locus studied.

When an equilibrium does not exist and the gene frequencies are changing over generations, as they most certainly did in the experiment described in Fig. 5.1, an estimate of the fitness values can be made from the observed curve formed by all the sample points. The values in Fig. 5.1 do not produce a perfectly smooth curve, since sampling error was involved, nor is the general trend linear, because the differences between the points (Δq's) appear less as the values of q become smaller. Such complications call for specialized techniques for analyzing the frequency changes and determining selection coefficients. Basically the procedures, like those worked out by Sewall Wright and H. Levene (see Spiess, 1962), consist of choosing (by iteration) adaptive values that, when substituted in Eq. 5.3, produce a theoretical curve of gene-frequency change that best approximates statistically the observed points (that is, the squared deviations between the theoretical and observed points are minimized). The values so obtained are considered "total" fitness values of the genotypes over the period described. The smooth curve in Fig. 5.1 represents the expected changes assuming adaptive values of the three genotypes to be $++ = 1.0$, $+i = 0.74$, and $ii = 0.63$, and that they were apparently constant throughout the study. These values indicate that there is a slight partial dominance of the i arrangement. Although the total adaptive values are similar to those found for the single fitness component, viability, we must suspect that components of fitness other than viability are contributing to the reproductive success or failure of the different genotypes. In fact, additional studies have revealed that there are also differences in egg-laying capacity and rates of development among the chromosomal types.

The three modes of selection

Selection within a population is described as *stabilizing, directional,* or *disruptive.* These terms refer to the discriminative acts of saving or rejecting phenotypes as the parents of the succeeding generation. However, they secondarily imply the kinds of changes that may occur in the underlying genetic variation. The response of a population to the force of selection depends not only on the modes and strengths of the selective forces but also on the action and interaction of the genes in the phenotypes being selected and on the variability and the genetic system of the population (Mather, 1955). Included in the last set of factors are the mating system, the size of the population, and the population's state of adaptedness. With respect to the second item, each form of selection may involve characters genetically influenced primarily by single genes, as discussed under the mathematical model, or it may involve quantitative variation controlled by the action and interplay of multiple genes.

The basic kinds of selection are illustrated diagrammatically in Fig. 5.4. They are depicted in terms of their directions of action on the existing group of phenotypes. The variation of a hypothetical

Fig. 5.4. The three basic modes of selection and the change in genetic variance expected from each.

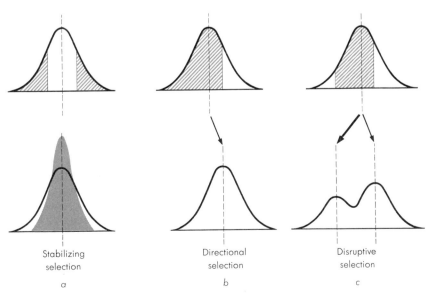

| Stabilizing selection | Directional selection | Disruptive selection |
| *a* | *b* | *c* |

metric trait is portrayed; in the frequency distribution of each group of phenotypes the majority of the individuals possess the character near the mean and fewer extreme deviates appear at either end of the scale. The sharp truncation of the selected group is an oversimplification. Such precise all-or-none culling is more descriptive of artificial selection. Natural selection usually involves differential reproduction throughout the total array and certain probabilities of survival (reproductiveness) for each phenotypic class. But the concepts are best presented by this simple approach. Again, it may be easier to visualize selection as differential survival, which occurs just before random mating (except in Fig. 5.4c, where mating may be partly assortative).

Stabilizing selection

A population always stands in relation to its environment. To exist, it must be adapted constantly to its habitat. Throughout the history of a population, selection incessantly acts to preserve those genotypes that direct the formation of phenotypes best suited for the particular set of conditions prevalent at any given time. If a population lives in a constant environment for several generations, it may have an opportunity to reach a high level of adaptedness (i.e., $\bar{W}$ is maximized with respect to the long-term effects of selection). If it does, the genetic variation will probably become stable and the gene frequencies at most loci will be near their equilibrium values (0, 1, or $\hat{q}$)—the population comes to rest on a "selective peak." But selection does not cease. It assumes a stabilizing role, since the tendency is then to maintain the status quo. This means that a certain assemblage of well-adapted genotypes (the adaptive norm) is preserved from generation to generation, while those forms deviating from the norm are more likely to be eliminated with each selection cycle. Selection continues to operate because poorly adapted forms are continually produced through one or more of the sources of variation—mutation, migration, and recombination and segregation. This elimination of the pronounced "phenodeviates" has been observed repeatedly in natural populations. These observations reveal that it is often the extreme forms, at either end of the scale of a metric trait, that are the most unfavored.

As a classic example, more than 60 years ago Bumpus studied the differential survival of a group of English sparrows which he found stunned and helpless during a severe snowstorm. Approximately half of the birds that he collected and took to the laboratory died, and the others survived. He made extensive measurements of several morphological features and discovered that the characters of the surviving sparrows conformed to the mean for the species (the adaptive norm) and that the birds that died showed a greater range of expression for the traits. In more recent times, Karn and Penrose recorded and grouped the birth weights of babies plus the survival rate from birth to the first 28 days of life for each weight class. They found that mortality was highest among the groups of infants weighing much less or much greater than 8 pounds at birth.

Another example of this type of selection, involving a character directly associated with reproductive rates, is Lack's study of clutch size (eggs per brood) in birds. He suggested that there is an optimum clutch size in each species, which has been so adjusted by natural selection as to yield the maximum number of offspring. Although the number of eggs produced obviously constitutes a fitness component, he discovered that clutches that are too large can be disadvantageous, because a negative correlation between the number of nestlings and the feeding ability of the parents exists; individuals in large broods receive less food and become liable to the hazards of later life. He reported that a brood size of 5 in the Swiss starling was optimum. Although the number of birds fledged from clutches of more than 5 eggs was higher, the mortality was greater among the less nourished birds after they left the nest.

Here we see how selection pressure for one character (survival) counterbalances another (egg production), producing an optimum condition (clutch size). This exemplifies the concept presented earlier that fitness is a composite of all aspects of the phenotype that contribute to the production of viable and fertile offspring. An individual and its phenotype and genotype, a population, and the gene pool are all highly complex systems, the separate parts of each of which have become mutually adjusted through natural selection. Natural selection leads to that balance among the components which results in the production of individuals and populations best adapted to the prevailing environment. Often, because of such balance, a single character is selected, not for its most extreme expression, but for some intermediate level.

This mode of action, in which a single optimum is favored consisting of the individuals at or near the mean of a quantitative character, is called stabilizing selection (Fig. 5.4a). A consequence of stabilizing selection is sometimes a reduction in the variance of the character, without a concomitant shift in the mean value (note the shaded curve in the figure). The variation narrows when selection is for the "average" but the population is not yet on a selective peak. It may also narrow if a population once in a balanced state is suddenly subjected to more intense stabilizing selection. However, when an equilibrium exists between the loss of deviates and their production, a stable pattern is expected over generations. In any case, the mean of the character is supposedly unaltered when the "ordinary" are continuously preferred.

Single-gene models of stabilizing selection are characterized by forces that maintain gene frequencies near their equilibrium values, 1 or $\hat{q}$. Alleles kept at intermediate levels ($\hat{q}$) are sometimes associated with balanced polymorphisms, in which the genotypic value of the heterozygote is adaptively superior (examples are given in Chapter 7).

Directional selection

Selection is opportunistic in that fitness values are defined by existing conditions, and adaptations are produced without a guarantee of worth to their possessors for the future. The longer a population remains in a stable habitat, the better its chance of reaching and maintaining a state of adaptedness. We should expect to find the alternative to a stabilized variation pattern, i.e., the most rapid evolution, when populations are exposed to novel environments. Rarer variants which were once relatively ill fit may then become the chosen type, and thus some phenotypic optimum other than the ancestral norm would be preferred. Since $\bar{W}$ is guided by selection toward its maximum amount, genes and genotypes responsible for such variants change progressively until a new balance is reached. When a single optimum is favored but it consists of phenotypes other than those making up the modal class, i.e., near one end of the phenotypic distribution, selection is called directional (Fig. 5.4b). The expected response is a change in the population mean toward the mean of the selected group. Reduced to the single-gene model, this is comparable to the cases in which one allele is favored and $\bar{W}$ has not reached its highest value. If we accept the fact that any directed change in the hereditary composition of a population constitutes a microevolutionary step, this concept, without genetic trappings, represents the core of Darwin's theory of the mechanism of evolution.

Examples of directional selection are too numerous even to list. You might note them as you see them in this text (in the next section) or in others (such as Dobzhansky, 1955).

Disruptive selection

The third component of selection can be viewed as the converse of stabilizing selection in that it concerns multiple forces acting on diverse forms of a single population. When selection favors more than one phenotypic optimum, i.e., quite different genotypes, and discriminates against intermediates between them, it is called disruptive. Two classes made up of the extreme types at either end of the phenotypic distribution are depicted as the selected optima in Fig. 5.4c.

The expected response to disruptive selection is an adjusted discontinuity of the variation pattern. The results can take one of two basic forms: (1) *polymorphism,* in which each optimum is represented by a distinct morph, or (2) *divergence and isolation,* in which the optima are composed of individuals in locally adapted "subpopulations." The primary condition necessary for the production of a polymorphism is that the classes selected are in some way functionally interdependent, whereas reproductive isolation is expected (in theory) with the disruptive selection of independent groups that generally become spatially oriented in a heterogeneous environment because of different ecological requirements (Mather, 1955). In either case, the outcome ultimately rests on *directional* effects, i.e., changes in the different selected classes take different directions, and a stabilizing tendency exists toward each optimum (Fig. 5.4c).[2] Depending on the

---

[2] In fact, where divergence among "subpopulations" is occurring, one could just as well analyze it in terms of *directional selection* in the separate demes, while also considering the amount of gene flow between them (see Chapter 4).

intensity of selection, the difference between the optimal phenotypes, and the persistence of the process over generations, the phenotypic distribution for a quantitative character should become multimodal with increased variance under disruptive selection. This expected response has been demonstrated in laboratory experiments with *Drosophila*, in which high and low bristle numbers were simultaneously selected under a prescribed random-mating program (Gibson and Thoday, 1964).

In nature, disruptive selection apparently has been responsible for the perfection of some polymorphisms. For example, a variety of mimetic females occur together in many of the geographical races of the African swallowtail, *Papilio dardanus*. Each of the several forms commonly found in the same locality resembles a different distasteful species of butterflies and so gains protection against birds, since the

Fig. 5.5. Nonmimicking and mimicking *Papilio dardanus* with their models. *Left,* female *(top)* and male *(bottom)* of *P. dardanus from Madagascar,* a nonmimicking race. *Top row, second,* a noxious model, *Danaus chrysippus,* from East and South Africa; *next,* the female of *P. dardanus* which mimics *D. chrysippus. Right,* an imperfect mimic that was a segregate from a hybrid between the mimicking race and a race not possessing this mimic. *Bottom row (starting with second figure),* a model belonging to another genus, *Amauris niavius domini-*

mimics themselves are relatively edible and subject to predation (Fig. 5.5). Many data are available which show that this type of mimicry is most successful when the mimics are substantially fewer than their models. The reason is that the predators must be repeatedly cued through experience to avoid the model. It would be difficult for a mimic to hide for long behind the guise of a distasteful form if that model were rare. It has been concluded therefore that a variety of mimics, such as that in *P. dardanus,* would be advantageous, since more than one model is copied. Presumably the polymorphism in this species has arisen through disruptive selection, the several distinct morphs (optima) being functionally related in that the success of each depends on the presence of the others.

Recent work, summarized by Sheppard (1961), has supported the idea that the major alleles controlling the various patterns were only

canus, its female that mimics *P. dardanus,* and an imperfect mimic that segregated from a hybrid with a nonmimicking race. Imperfect mimics are produced by the action of modifier complexes occurring in regions where models are infrequent. From G. Ledyard Stebbins, *Processes of Organic Evolution.* Englewood Cliffs, N.J.: Prentice-Hall, Inc., 1966; after P. M. Sheppard, *Cold Spring Harbor Symposia Quant. Biol.,* **24** (1959).

weakly effective at their inception and that each mimetic pattern has subsequently been perfected through the selection of modifiers particular to each race in which that mimic is found. Moreover, it is known that dominance is expressed among the multiple series of alleles occurring in each population. When races are crossed, and the major genes from different gene pools are combined in a heterozygote, however, intermediate phenotypes (which of course are poorer mimics) are formed. This absence of dominance between allopatric forms suggests that the dominance expressed among codemic types results from particular genetic backgrounds (modifiers) that have evolved in each race.

In discussing fitness and selection models, we implied the simplifying assumption of a uniform environment in which adaptive values were constant over a given selection period. This is seldom so in nature. Environments vary both in time and space, and selective forces change with respect to both. Different species adjust to variations in environment in different ways: (1) Some become highly specialized; these are usually rare forms that are restricted in range to a uniform and stabilized area. (2) Some have incorporated genotypes with wide reaction norms; through phenotypic flexibility the individuals adjust developmentally and physiologically to changes in their environment that may occur at various times. (3) The example discussed above is typical of those species that extend their utilization of a heterogeneous environment through genotypic flexibility (polymorphism) by the addition of distinct morphs, each adapted to a particular subniches in a local area. (4) Animal species that are sessile or exhibit low vagility and most plants are usually precisely adapted to their immediate environments. Usually, in addition to genetic and phenotypic flexibility, many of these species adjust by breaking up into groups (ecotypes), each of which is adapted to the specifically defined ecological situation within one or more subareas of the species' range. The "subpopulations" are usually contiguous but maintain their integrity, through ecologically specified selection pressures, despite gene flow from the other neighboring "ecological" races. (5) Species with high vagility, such as many higher animals, usually have a more extended range, meeting many sorts of physical and biotic environments. Some of these respond to the environmental variation by forming geographic races, each occupying a relatively more extended portion of the species' range (polytypism). The difference between (4) and (5) is mostly a matter of degree (ecotypes are at least microgeographical units).

The kinds of responses listed under (3) and (4), i.e., ecological polymorphism and divergence, are presumably brought about by disruptive selection, since in each case multiple optima exist together with some degree of interbreeding between the selected groups. However, there is considerable disagreement among population geneticists on whether the divergence among "subpopulations" can include genetic isolation (sympatric speciation) in the face of gene exchange among the groups. Mather (1955) has proposed that "where variation in the environment results in different optima being favored in different groups of the population, the resulting tendency will be . . . toward genetic isolation of these groups one from another." For this tendency to be realized, however, ". . . the optimal phenotypes must be independent of each other . . . the groups must be sufficiently distinct . . . [and] . . . the separation of the populations must persist in time." That very strong disruptive forces can lead to isolation among genotypes within a population is theoretically possible and has actually been demonstrated in laboratory experiments with *Drosophila* (Thoday and Gibson, 1962). Mayr (1963), however, argues that the intensity of selection needed for such separation is far greater than that ever expected to occur in nature. Although there is a general consensus that disruptive selection can lead to phenotypic polymorphism ([3] above), including genetic diversity associated with environments that are heterogeneous in space ([4] above), Mayr suggests that it can never result in sympatric speciation in view of changing selection pressures and gene flow from contiguous demes. The controversy seems to resolve to the question of just how much gene flow is necessary to prevent the genetic amalgamation of ecologically distinct subgroups of a population. We have as yet no pat answer.

Selection in natural populations

Examples of progressive selection fall into four categories: first, the continued improvements in yield; chemical content; disease resistance; and meat, milk, and egg production made through artificial selection in the breeding of agricultural crops and stocks are most familiar demonstrations. These results, more than any other evidence, point out the tremendous power of selection in altering the characteristics of a species. Second, many artificial and natural selection studies have been carried out in the laboratory. The *mulleri* experiment presented earlier is one illustration. The major purposes of this sort of investigation have been to test certain theoretical models and to procure additional basic knowledge on which to build more perfected models (see Falconer [1960] for a review concerning quantitative inheritance). The myriad and fascinating ways organisms meet the challenge of their environs through specific adaptations represent the third type of evidence. It is speculated that adaptive characteristics, although maintained by stabilizing factors, originally are incorporated into a species' "norm" by directional selection much in the same way that progressive changes lead to economically important improvements in cultigens (field crops and domestic animals). Although this evidence is inferential, the results of selection being observed a posteriori, it is highly complemented by the final group of examples—the direct observations of secular changes in natural populations. Nearly all of the latter studies have involved polymorphisms in populations recently exposed to new conditions.

By far the most thoroughly analyzed and spectacular changes

recorded concern the increase of melanic forms in moths (reviewed by Ford, 1964). This phenomenon has been dubbed "industrial melanism," because the gene-frequency changes are associated in time and location with the spread of industrialization during the past several decades. It has taken place around the industrial centers of northern Europe, the United States, and particularly in England, where more than 70 species of moths are known to have melanic forms. The most extensive studies have been carried out by British workers with the peppered moth, *Biston betularia* (L.). All specimens of this species collected before the middle of the nineteenth century were grayish white with black mottling over the body and wings. This *typical* color pattern (*typical* being the name given to gray forms) seems to confer a protective camouflage to the insect when it is at rest during the day on lichen-covered tree trunks. Many present-day *betularia* populations are polymorphic and include a totally black morph, *carbonaria*. The polymorphism is controlled by alleles at a single locus, *carbonaria* being dominant to *typical*. A single such melanic specimen was first collected in Manchester, England, in 1848. It is estimated that the dark form represented only about 1 percent of the population at that time. Its frequency increased rapidly, however, in Manchester and in other big cities until, by the turn of the century, it had become the common form (more than 90 percent in several populations). A most pertinent discovery was that the distribution of these populations is noticeably correlated with the urban manufacturing districts of England. In fact, populations with a majority of *typical* individuals are found today only in Ireland, the north of Scotland, and the southwest of England.

Kettlewell (see summary of work in Ford, 1964) has investigated the factors responsible for the obvious advantage of the melanic form in the industrial areas. He found that the fantastic speed at which it replaced the paler type resulted from strong directional selection, based primarily on differential predation by birds. His beginning thesis was constructed on the observation that lichens are absent in the smoke-polluted woods around factory-filled urban communities and that the trees have become blackened with soot. In these places the black form of *betularia* is cryptically colored, while the speckled type is less conspicuous on the dappled-gray background afforded by lichens in the unpolluted rural countryside (Fig. 5.6).

He chose two areas for study, a polluted woods near the city of Birmingham, where *carbonaria* represents nearly 87 percent of the population, and an unpolluted region in Dorset, occupied by a monomorphic population of *typical*. A series of mark-release-recapture experiments were conducted at each site. A certain number of marked individuals of both types were released, and the percentage of each recovered later was recorded. The figures supported the hypothesis that the different morphs are selectively eliminated; the moths that contrasted with their background were recaptured proportionately less than their better-camouflaged counterparts were. For instance, during one test, 154 *carbonaria* and 64 *typical* moths were released in the

Fig. 5.6. The *typical* and *carbonaria* forms of *Biston betularia,* the peppered moth, are shown at rest on a lichen-covered tree trunk in the unpolluted countryside on the left and at rest on a soot-covered oak trunk near Birmingham, England, on the right. The *typical* form may be found in the left-hand picture just below and to the right of the black moth. Courtesy of H. B. D. Kettlewell.

polluted Birmingham woods. Of these marked individuals, 82 *carbonaria* (52 percent) and 16 *typical* (25 percent) were recaptured. Conversely, of the 406 *carbonaria* and 393 *typical* forms set free in the unpolluted habitat in Dorset, 19 black and 54 pale individuals were recaught, being 4.7 and 13.7 percent, respectively. The complementary results obtained in the different habitats show that selection indeed occurred and that the more protectively colored individuals in each environment were preferred.

One selective agent militating against the conspicuous types in each case was discovered by direct observation (Kettlewell and Tinbergen). Equal numbers of the dark and light moths were placed on tree trunks and then watched and photographed throughout the day from a blind. Soon birds were seen searching for, taking, and eating the moths. At Birmingham, 43 of the easily detected *typical* forms to 15 of the black *carbonaria* were eaten by redstarts. Opposite results were found in the unpolluted forest in Dorset; 5 species of birds devoured 164 *carbonaria* and only 26 *typicals,* the latter usually being eaten only after the more conspicuous black moths were completely eliminated.

This simple story is complicated slightly by the following:

First, a third morph, *insularia*, is known in the peppered moth. Its appearance is somewhat intermediate between *typical* and *carbonaria*. Like *carbonaria*, this pattern is dominant to *typical*, but it is controlled by an allele at a different unlinked locus. The frequency of *insularia*—separable only from *typical*, since *carbonaria* is epistatic—is seldom high. It is most commonly found in areas where pollution is marginal. This again supports the hypothesis that crypsis is correlated with the amount of industrial debris, the intermediate phenotypes being favored in the intermediate environments.

Second, as one might suspect, the alleles at the *carbonaria* locus are apparently pleiotropic. They affect not only the color patterns but the behavior and relative viability of the moths. Kettlewell further found that individuals of each morph tend to come to rest on non-contrasting surfaces when a choice is available. Also, there is some indication that when mixed broods are raised in the laboratory, the melanic forms possess a physiological advantage during the immature stages, especially if the larvae are starved. There is some doubt concerning this point, however, since studies of backcross broods made in 1900 showed no significant viability differences between the color types. It has been suggested that the survival superiority of the melanics is real enough but that it has just recently evolved. Perhaps as the melanic individuals increased in the populations, the total gene complex of the species was reconstructed, producing a new genetic milieu (modifier genes) that enhanced the behavior and viability of the ecologically superior *carbonaria*. This explanation is also given for the next fact concerning the peppered moth.

Third, the dominance of the *carbonaria* gene has evidently evolved during the past one hundred years. The first melanics captured, although much darker than *insularia*, were not nearly so black as present-day *carbonaria* heterozygotes, which are indistinguishable from the homozygotes. It is safe to say that the early types were undoubtedly heterozygotes, because the frequency of the mutant allele was so low. If, as speculated, the early homozygotes were darker than the heterozygotes, then dominance has indeed evolved between the alternate alleles.

Industrial melanism is an example of directional selection. A gene that once occurred as a rare mutant suddenly became advantageous, spreading through the population at the expense of its former normal allele and reducing the latter to the mutant status. During the transition from rarity to prevalence, both the "mutant" and "normal" forms were present in goodly numbers in the species, thereby manifesting a polymorphism. Genetic polymorphism has been defined by Ford as the occurrence together in the same population of two or more distinct forms of a species in such proportions that the rarest of them cannot be maintained merely by recurrent mutation. Because of the temporary nature of gene replacement, however, processes like industrial melanism are classified as *transient* polymorphisms (compare *balanced* polymorphism, described earlier in this chapter.)

The interaction of natural selection and other evolutionary modes

Each of the basic evolutionary forces—mutation, migration, selection, and genetic drift—which are the elementary agencies known to be responsible for changing gene frequencies, have now been singly quantified by the use of simple models. We have also distinguished two aspects of the evolutionary process: the generation of the raw materials of evolution (allelic and genotypic diversity) and the reshaping of the genetic variation so produced. Mutation and migration, along with recombination intrinsic in the process of sexual reproduction, give rise to the genetic variation. However, only a portion of the genotypes engendered are represented among the progenitors of the genetic continuum of the population—the rest are lost. We have just seen that such loss in the second stage of the process is discriminative when the continuum (alternating gametic and zygotic phases) passes through the sieve-type screening process of selection, and we previously learned that elimination may also occur randomly by accidents of sampling, or genetic drift, in finite populations.

At various times in the development of evolutionary theory, one or another of the basic modes has been assigned the supreme role as the "directing" or "causal" agent of evolutionary trends. There were periods dominated by selectionists, mutationists, and even "drifters." However, the present-day synthetic theory, which is almost universally accepted, recognizes the intimate relation of both aspects of the process and the obvious interplay of the various factors for the realization of adaptation.

Advocates of modern neo-Darwinism still purport, however, that natural selection is the "guiding" agent of evolutionary progress, since it is the only force oriented with reference to a standard of adaptedness. When and what type of variation arises in a population or becomes lost by genetic drift is mainly a matter of chance. For example, mutation and recombination if left unchecked would produce a diverse array of aberrant variants and malformations, since the great majority of novel and reoccurring mutants are detrimental and since the process of recombination (sexual and parasexual processes) simply reassembles gene combinations and does so without respect to the needs of the organism. It is extremely improbable that the highly integrated genotypes responsible for the perfected designs of major adaptations, such as the vertebrate eye or the strobilus of a conifer, could have arisen by the fortuitous combinations of chance variations.

Natural selection has replaced the more archaic philosophies of autogenesis and finalism as the force believed to give directional guidance—observed a posteriori—to evolution. As aptly stated by Sir Ronald Fisher (1954), natural selection is the ". . . process by which contingencies *a priori* improbable, are given, in the process of time, an increasing probability, until it is their nonoccurrence rather than their occurrence which becomes highly improbable." Adaptations originate through natural selection. It is argued that selection reduces the diversity of organisms, but it does so discriminately—it saves as well as rejects genotypes; and, by accumulating adaptive alleles and gene combinations through time, it can be said that its action "originates" organic form and function. A classic metaphor likens selection to the creative art of a sculptor, who reveals a statue, cryptically imbedded in a chunk of marble, by chipping away the unwanted pieces.

References

Dobzhansky, Theodosius, *Evolution, Genetics and Man.* New York: John Wiley & Sons, 1955.

Falconer, D. S., *Introduction to Quantitative Genetics.* New York: The Ronald Press Company, 1960.

Fisher, R. A., "Retrospect of the Criticisms of the Theory of Natural Selection," in *Evolution as a Process*, Julian Huxley, A. C. Hardy, and E. B. Ford, eds. (London: George Allen and Unwin, Ltd., 1954), pp. 84–98.

Fisher, Ronald A., *The Genetical Theory of Natural Selection.* London: Oxford University Press (Clarendon), 1930. This classic work is now available in a paperback edition (New York: Dover Publications, Inc., 1958).

Ford, E. B., *Ecological Genetics.* New York: John Wiley & Sons, 1964. A most thorough review of the research on the genetics and ecology of natural populations by British workers. The phenomena of drift, natural selection, and polymorphism are particularly well covered.

Lerner, I. M., *The Genetic Basis of Selection.* New York: John Wiley & Sons, 1958. An advanced treatment of selection.

Mather, Kenneth, "Polymorphism as an Outcome of Disruptive Selection," *Evolution, 9* (1955), 52–61.

Mayr, Ernst, *Animal Species and Evolution.* Cambridge, Mass.: Harvard University Press, 1963. See Chapter 8.

Sheppard, P. M. "Some Contributions to Population Genetics Resulting from the Study of the Lepidoptera," in *Advances in Genetics*, Vol. 10, E. W. Caspari and J. M. Thoday, eds. (New York: Academic Press, Inc., 1961), pp. 165–216.

Spiess, Eliot B. *Papers on Animal Population Genetics.* Boston: Little, Brown & Company, 1962. The papers by the following authors are of special interest: J. B. S. Haldane, p. 3; S. Wright and T. Dobzhansky, p. 68; C. C. Li, p. 123; T. Dobzhansky and H. Levene, p. 139; T. Dobzhansky and O. Pavlovsky, p. 421; R. C. Lewontin and L. C. Dunn, p. 442; and H. B. D. Kettlewell, p. 477.

Thoday, J. M. and J. B. Gibson, "Isolation by Disruptive Selection," *Nature 193* (1962), 1164–66.

Wallace, Bruce, "The Comparison of Observed and Calculated Zygotic Distributions," *Evolution, 12* (1958), 113–115.

Six

Genetic Drift

Natural selection is the foremost factor responsible for significant evolutionary change. We have said previously that a certain erosion of genetic variation characteristically accompanies a trend toward adaptedness. A second agent known to reduce hereditary variation in populations is genetic drift (see Chapter 3). Genetic drift involves chance fluctuations in allele frequencies due to sampling error, and there is a tendency toward fixation of one or another allele, especially in very small populations. Since accidents of sampling are mathematically defined certainties, given the Mendelian mechanism of inheritance, genetic drift must be considered an omnipresent and undeniable agent in finite, and thus in all, natural populations. However, the evolutionary significance of such random changes has been a perennial subject of debate.

The interaction of random and nonrandom factors

The theoretical considerations of random processes, developed primarily through the work of Wright, have been verified experimentally in both living and simulated populations. The results from the computer studies by Schaffer, which most dramatically demonstrate the dispersal process, were known to be due completely to random processes, since selection, mutation, and migration were purposely omitted from the model (reexamine Fig. 3.7).

Although migration and mutation could be ignored in a group of comparable experiments by Kerr and Wright, using real populations (*Drosophila melanogaster*), it was not possible to control selection. Kerr and Wright kept 96 separate small populations by taking a random sample of 8 individuals (4 ♀ and 4 ♂), each generation to be parents for the next. Initially each "line" possessed the gene for "forked bristle" and its normal allelomorph, each at a frequency of ½. After 16 generations, the forked and wild-type alleles were still segregating in only 26 lines, while the forked gene had become fixed in 29 and the wild-type allele was fixed in the other 41. Here it is shown that genetic drift tends to produce homozygosity and that chance primarily determines which allele is fixed in a given population. However, the results were apparently confounded by selection, which seemed to favor slightly the wild-type allele. The effects of selection were even more obvious in another series of populations involving the *Bar* gene, the wild-type allele being fixed in 95 of 108 lines, while the *Bar* allele reached 100 percent in only 3 lines. The point is, however, that fixation occurred in the majority of the lines, and sometimes the seemingly superior allele was the one lost. Such an outcome (random fixation of ill-adapted traits) would not be expected in a large population in which selection predominates. Kerr and Wright also showed that the occasion for fixation by drift is drastically reduced when selection strongly favors heterozygotes. In a third series of small populations carrying alleles producing heterosis (the heterozygous genotypes were superior to the homozygotes in fitness-component tests), there was a dispersion in gene frequencies among the population, but just 8 of 113 lines became genetically homogeneous within 10 generations.

These experiments demonstrate that random gene-frequency fluctuations indeed take place in small isolated populations, producing results suggested by the mathematical theory. That the theory is correct is not doubted. The issue is whether or not genetic drift has any evolutionary significance. We have seen that mutation, migration, and recombination are factors that intrinsically play important roles in the origin of variation and hence through natural selection to the production of adaptation. The weight of drift is less clear. Drift is a biologically unoriented process similar to mutation or recombination, although it has to do with the loss of variation rather than with its generation. It is a companion factor to selection in the reshaping of existing variation, but unlike selection, it is indeterminate and tends only to cause disorder in nature. Apparently, any positive role it may play must in some way be tied up with its relation with the factors causing systematic genetic changes, particularly by enhancing the fixation of favorable alleles and gene combinations or by counteracting the conservative nature of stabilizing selection. Even the strongest supporters of drift as a significant evolutionary mode are quick to point out that it is only an adjunct to natural selection.

The problem as posed above is in no way simple, since it requires a consideration of the interaction of all evolutionary forces from a quantitative standpoint. Although genetic drift depends per se on the effective breeding size of the population (N), its effects are conditioned by the extent and direction of the systematic pressures, mutation (u), gene flow (M), and especially selection (s). From Wright's calculations, it is evident that random fixation is more likely when N, M, s, and u are all low, i.e., in small, isolated populations in which the alleles involved are selectively neutral and are seldom replenished by mutation once they are lost. However, "small" is a relative term; it must be quantified with some cognizance of s, M, and u, as well as N. This is exactly what Wright has done. While his formulations are rather abstruse and far beyond the scope of this text, his basic conclusions seem obvious. The smaller the effective size of the population, the greater is the random fluctuation in gene frequencies; while at the same time, the stronger the systematic factors, the less obvious becomes the indeterminate process. Treating each directed force separately in relation to its coaction with drift, we can state more quantitatively that a population is "relatively" small (i.e., drift predominates) when $4Ns$, $4Nu$, or $4NM$ is less than unity and that a population is "large" (genetic changes are mostly determinate) when one of these products is greater than one. Thus, a closed population of a quarter of a million individuals is small for a neutral allele with the typical mutation rate of 10^{-5}, while another one consisting of 250 individuals is very large, say when the coefficient of selection is more than 1 percent.

Although alleles that have drifted to extinction may be replenished by mutation, recurrent mutation pressure may be ignored as a force significantly great enough to oppose drift in very small populations, because mutations usually occur with such low frequencies. However, this may not be true of many isoalleles recently revealed through the use of electrophoresis. There is some evidence that the mutation rates of isoalleles are much greater than those commonly found for so-called major genes. For the present, however, we shall consider that genetic variation can hardly be maintained by the process of mutation in the face of drift in small populations. This is not true for selection or gene flow because these forces may occur with much greater intensities. What would be regarded as a low rate of migration (i.e., $M = 1/4N$, or 1 immigrant per 4 generations, regardless of the value of N) is sufficient in reducing drift to a minor role in changing gene frequencies. Selection intensities as low as 0.01 would suffice in checking the random aspect of fixation in all natural populations except the very smallest (those with less than 50 individuals).

Studies of natural populations

The obvious question to ensue is whether any phenomena in nature have occurred primarily as a consequence of genetic drift. This explanation has been suggested for several, but every example has been controversial, because any evolutionary change for which drift is suspected to be responsible could nearly always as well be explained by varying selection pressures differing in either time or space. In order to invoke drift as a causal agency, it would first be necessary to disallow the importance of selection. This is hardly possible, for no matter how adaptively neutral a particular trait appears, the genes involved may very well have profound effects on some unmeasured fitness component. But where the proof of this is unavailable and the other conditions seem appropriate, drift may be a reasonable explanation.

Genetic drift is a random process of change which may be expressed in one of two ways: (1) differences between successive generations of a single population observed over time, or (2) differences among a contemporaneous group of related populations observed at one time over space. For practical reasons most studies have been concerned with comparisons of the latter type, although such evidence would have to be considered indirect, since it was observed after the fact. Since gene-frequency dispersion is cumulative and relatively permanent, drift may occur, not only in populations that remain small, but also in those that become small periodically (here drift equals the "bottleneck effect") or in those that are small at their inception, i.e., those established by a few emigrants, or founders, carrying a small sample of genetic variation from a larger population (here drift equals the "founder principle").

Some of the most convincing illustrations of probable random genetic changes have been reported in man. From earliest history mankind has been subdivided into many populations of various sizes, being more or less isolated by geographical, political, religious, and social barriers. Although a "population explosion" has been taking place and man now occupies nearly all parts of the earth, there still exist several tribal groups and isolated religious sects with small effective breeding sizes. Even some of our larger populations are believed to have arisen in the not-too-distant past from a few original founders. Nearly all the populations of man have been extensively surveyed genetically for the various blood-group polymorphisms, and it has been found that the inherited characteristics of the small restricted populations often differ strikingly from adjacent or surrounding populations.

Most human populations are polymorphic for the ABO blood groups, possessing among their members all three alleles, I^A, I^B, and I^O. The inhabitants of different regions of the earth can be characterized as "serological races" on the basis of the relative frequencies of these alleles. Usually such racial differences are only a matter of degree, since most races include individuals belonging to each of the four blood groups, O, A, B, and AB. However, South American Indian tribes are, or were before the European influx, isogenic for group O; the I^A and I^B alleles are rare in most North American Indians as well. The Blood and Blackfeet Indians of southern Canada and Montana

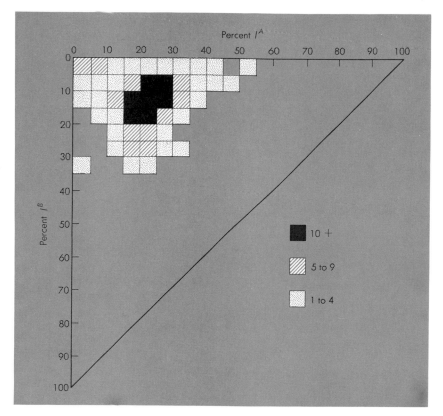

Fig. 6.1. The distribution of **ABO** allele frequencies in 215 representative human populations from throughout the world. The entire triangle encompasses the total range of possible frequencies of I^A, I^B, and I^O, in which the percent of I^O equals 1 minus the percent of $I^A + I^B$. It can be seen that the populations are grouped within a small area of the entire range. Redrawn from Curt Stern, *Human Genetics*, after the work of Alice M. Brues in *Am. J. Phys. Anthropol.*, 12 (1954), 559–97.

are exceptional: unlike the tribes to the south, they carry the I^A allele, and unlike nearly every other human population, this allele occurs in very high frequencies, up to 80 percent. The Amerindian stocks were presumably derived from Asian Mongoloids, a small number of whom perhaps migrated across the Bering Strait. Although every blood group occurs among the Orientals, group O being relatively low, opportunities for drift may have existed among these founders or in the small isolates formed later, and chance alone, or in conjunction with selection, may have produced these unusual frequencies.

The I^A allele ranges in frequency from 15 to 30 percent in the majority of human populations, while the I^B allele usually makes up 5 to 20 percent of the total (Fig. 6.1). The larger Eskimo populations from Alaska across the continent to Greenland and Labrador, for example, have frequencies of about 30 and 6 percent respectively, for the

I^A and I^B alleles. However, a small tribe of Polar Eskimos in Thule, Greenland, numbering less than 300 individuals and known to have had little recent gene flow from other populations, has only 9 percent I^A genes. Moreover, the I^B gene has become completely lost in some other small tribes in Labrador and Baffin Island. Again drift is suspected.

Glass and coworkers (1952) carried out a most interesting study of a Dunker community in Pennsylvania. The Dunkers are a religious sect that immigrated from Germany in the early eighteenth century and have since remained in small communities, intermarrying very little with their surrounding countrymen. It might be expected that the blood-group proportions of the Dunkers would lie somewhere between those of their eastern American neighbors and their Rhineland German brothers. This was not found to be true. Whereas blood group A constituted about 40 and 45 percent of the non-Dunker American and German populations (sampled in 1939), it had increased to nearly 60 percent in the Dunkers ($I^A = 0.38$), and the I^B allele had decreased to a near-extinction level ($I^B = 0.025$). The frequency of the L^M allele of the MN series was similar in the West German and United States populations, $L^M = 0.54$, but was considerably higher among the Dunkers, $L^M = 0.65$. These results, along with additional data on a few other traits, indicate that drift occurred in the small group during the 200-year period of self-imposed isolation, or among the 27 original founding families (see Glass, 1954). Evidence for the former interpretation, that some genetic changes have taken place within the community since its origin, was found by Glass in a further study. He grouped the individuals of the population by age into 3 successive "generations," 3 to 27, 28 to 55, and 55 or more years old. Although no significant differences of the ABO or Rh blood groups were found between the age groups, the gene frequencies of the MN series varied as follows: they were 0.55 L^M and 0.45 L^N in the oldest generation, being identical with the frequencies found in the German and United States populations; they were 0.66 L^M and 0.34 L^N in the middle generation; and they diverged further in the youngest generation to 0.735 L^M and 0.265 L^N. As pointed out earlier, this type of data is direct proof that changes have in fact occurred. However, the causes of the changes must be determined from other clues.

Many other examples are available of striking differences in gene frequencies among small related tribal groups or between social isolates and their more numerous neighbors in the same area. That isolation is an important requirement for such differentiation is shown by the contrasting blood-group distribution found among the larger populations of man. Adjacent large populations with few restrictions on migration have similar numbers of each group. Where gene-frequency changes occur, they do so gradually from one population to the next, forming clinal patterns of variation. For instance, the frequency of the I^B gene is high among the peoples of northern India and central Asia and exhibits descending gradients radiating toward the West, across Europe (see Fig. **7.2** in Victor A. McKusick's *Human Genetics*, in this series), toward the Northeast, through China and by way of Alaska to the Americas, and toward the Southeast to Australia. There are several centers of high I^A frequencies (e.g., northwestern Europe, Australia, and the upper regions of North America—remember the Blackfeet Indians), and multiple clines go in various directions. Thus gene flow among the larger populations seems to overcome any tendency toward an abrupt differentiation as observed in the small isolates.

It seems reasonable that the genetic diversity of the small groups of human isolates has come about by sampling accidents. Actually, the world-wide distribution of the blood groups in general has been championed for many years as an ideal example of nonadaptive variation, because it was thought that the genes concerned were selectively neutral. This view is changing. Some time ago Fisher pointed out on theoretical grounds that few genes can reach or maintain appreciable frequencies without some selective value. This would be particularly true of major genes associated with long-standing polymorphisms. That the ABO polymorphism is old is revealed by the fact that the various antigens are found throughout the human species. And, since similar or identical antigens have been discovered in several other primate species, one may conclude that the polymorphism is a part of man's prehuman heritage.

Ford has long espoused the concept, based on ample evidence, that polymorphisms are controlled by major genes (or switch-control units) with significant pleiotropic actions and that their development and maintenance results from a balance of selection forces acting on all the morphological and physiological characters so produced. "The existence of polymorphism therefore always advertises a situation of importance . . ." (Ford, 1964). In the early 1940's he predicted that the blood-group polymorphisms were balanced and that the genes involved owed their selective importance to an influence on fertility, viability, or susceptibility to disease.

Since then, the well-known hemolytic disease of newborns, erythroblastosis fetalis, has been thoroughly described in relation to mother-child Rh incompatibility (a concise review is given in Stern's *Principles of Human Genetics*). In short, the condition occurs among Rh-positive newborns of Rh-negative mothers who have become sensitized by the Rh antigen from the fetus, or by previous Rh-positive pregnancies. The blood cells of the fetus contain the Rh antigen. If the antigen gets into the blood stream of the mother (through the placenta), it induces the formation of Rh antibodies in the plasma of the mother (isoimmunization), which in turn leads to the destruction of the red cells of the fetus (hemolysis), resulting in anemia (see Fig. 5.1 in McKusick's *Human Genetics*).

Similarly, maternochild incompatibility from antigenic differences in the ABO system sometimes gives rise to erythroblastosis of the newborn or, more often, to early embryonic abortion. An ABO incompatible combination is one in which the mother carries those anti-A or anti-B antibodies for which the corresponding antigens are present in the cells of the fetus. Since group O mothers possess both types of antibodies, they run a greater risk of fetal loss than do mothers of non-O types, particularly those belonging to group AB. After reviewing the basic inheritance and serological mechanisms of the blood-group systems, the reader should enumerate potentially unfavorable marriages and demonstrate that intrauterine selection invariably discriminates against heterozygotes. This being so, it follows that the polymorphisms cannot be maintained by these selection forces alone, since selection against heterozygotes theoretically leads to an unstable condition favoring the fixation of the most common allele.

Other diseases have been linked with the ABO system. It seems highly probable that people belonging to group A are more prone to carcinoma of the stomach, carcinoma of the female genitalia, pernicious anemia, and diabetes mellitus, while compared with the normal population, group O occurs in excess among sufferers of gastric and duodenal ulcers. But in no case has a heterozygote been shown to possess a selective advantage, nor have the ABO disease associations been directly tied up with any racial differentiation. However, it is becoming more and more evident, as predicted by Ford, that the divergence of the blood-group frequencies cannot be explained entirely by random fluctuations. If drift were the only factor involved, for instance, we would not expect to find the frequencies of the many populations studied clustered within one-fifth of the total range of possible frequencies (Fig. 6.1). Also, Mourant (1959) has pointed out that the allele frequencies of the ABO system vary more from one population to the next than do those of the MN and Rh systems, indicating that the former are subject to more rapid changes by the features of the environment than the others are.

Genetic drift has been called upon to explain many sorts of differentiations among various animal and plant populations, especially where the populations are small and widely separated (dependent upon the vagility of the forms concerned) from other related populations. But only in a few instances has any real attempt been made to describe the possible selection forces involved. A most illustrative case of supposed genetic drift which *has* been extensively investigated concerns the color polymorphism in the land snail, *Cepaea nemoralis,* studied by Maxime Lamotte in France and by A. J. Cain and P. M. Sheppard in England. These snails have a discontinuous distribution existing in limited populations, or "colonies," in various habitats, such as woods, hedgerows, tall grasses, and open meadows, more or less separated by unfavorable intervening regions. A striking color and pattern variation exists in nearly every population. Individuals have brown, pink, or yellow shells. Some are self-colored, while others are striped, possessing one to five dark longitudinal bands. The color differences are controlled by a multiple series of alleles, yellow (y) being recessive to pink (Y)

and brown (Y^b) being dominant to both Y and y. The unbanded condition (B) is dominant to all forms of banding (b), while other alleles control the number and position of the stripes. Cain and Sheppard have shown that the two loci, Y and B, are tightly linked, constituting a "supergene" in inheritance. The color and striping polymorphism is known to have been a rather constant feature of these snails for a large part of their history. In fact, the different banding types have been found among fossil shells from Pleistocene deposits in England and France and in frequencies not unlike those in the populations now living in the same regions.

Lamotte has made an extensive survey of the distribution of *Cepaea* throughout France. He reports that the average frequencies of both the main banding patterns and the color types vary greatly in the different districts, while a considerable colony-to-colony diversity is additionally found within each district. This divergence among populations, regardless of the primary factors responsible, is obviously enhanced by the restriction of gene flow due to the low mobility so characteristic of the species. Since *Cepaea* occupies many types of habitats, it may be that the colonial diversity has resulted from varying selection forces associated with the natures of the habitats concerned. On the other hand, the differences could be explained by random gene-frequency fluctuations, in which case they should be greater the smaller the colony, if we assume along with Lamotte that natural selection operates independently of population size. He recorded the frequencies of the yellow and bandless genes in several hundred colonies of various sizes and found that the small populations did vary more and included more cases of homoallelism than the large ones did; moreover, the frequencies among neighboring small colonies were quite unrelated (Fig. 6.2), while those of juxtaposed large populations were somewhat more similar. These observations definitely show that the degree of differentiation is dependent on population size. Without necessarily denying the existence of environmentally imposed selection forces, Lamotte (1959) ascribes to genetic drift the increase in diversity among the small populations.

Cain and Sheppard argue that our initial assumption—the effects of selection being independent of population size—is not appropriate, at least in the case of snails, however, since the environmental backgrounds differ considerably over short distances. Their thesis is that the larger populations will be dispersed over more extensive areas and will encounter a greater number of habitats than the smaller ones will; thus, larger colonies will be more nearly alike than the smaller, because each will be adjusted to an average of diverse conditions (averages tend to be alike), while the latter will be specifically adapted to particular, locally homogeneous, habitats.

Evidence in support of Cain and Sheppard's thesis was found in studies of *Cepaea* populations near Oxford, England. In addition to demonstrating the typical colony-to-colony variation, they found a

Bandless snails

Banded snails

Kilometers
0   5   10

Pamiers

Ariège River

Ax

L'Hospitalet

a

b

c

Fig. 6.2. The relative proportions of bandless (a) versus banded (b and c are two of several types) snails in a series of adjacent colonies along the Ariege River in the Pyrenees Mountains of France. Note that the frequencies change somewhat haphazardly from one colony to the next. From Verne Grant, *The Origin of Adaptations* (New York: Columbia University Press, 1963), after the work of M. Lamotte, *Bull. Biol. France Belg., Suppl.* (1915), p. 35.

connection between the phenotypic composition of the populations and the kinds of habitats (Fig. 6.3); the commonest varieties were those least conspicuous in the habitat. The yellow shell color was found to be common in the green milieu of hedgerows and grasses, while brown snails were predominant in woods, where decaying leaves and bare ground have a dark appearance. The banded condition was more abundant in habitats providing a backdrop of many contrasts, such as hedgerows and rough, while the unbanded types were associated with uniform backgrounds. These correlations are good evidence of natural selection in relation to the environment. One can easily guess that this is another case of predator-oriented protective coloration. It was found that the song thrush breaks open the shells on suitable stones ("anvils") and eats the body of the snails. The selective nature of the predation was demonstrated by comparing the frequencies of the various patterns among the broken shells, accumulated around the "anvils," with those in the living populations in the same area. Invariably, the color varieties judged to be the most conspicuous under the prevailing conditions were represented in greater abundance among the broken shells. Release-recapture data have also added to the stockpile of evidence for selective predation. But, although visual selection apparently adjusts the relative frequencies of the various forms in each habitat, the forces responsible for the balanced nature of the polymorphism have not been determined—just as with the human blood-group polymorphisms, heterozygote superiority has not been shown, although there is reason to believe that heterosis is associated with the physiological traits that are simultaneously controlled by these genes. Although there are other ways in which balanced polymorphism may be maintained (Chapter 7), such as gene flow between populations adapted to different habitats, none seems to supply a sufficient explanation in this case. Finally, we must recognize that the correlations are not perfect, and as yet, it is still reasonable to think that drift, possibly in the form of the founder principle, has been instrumental in producing some of the diversity in the smaller colonies.

In discussing the foregoing examples we may have given the impression that one must work from an "either-or" antithesis, in which drift and selection are seen as alternatives, one or the other being solely responsible for an observed genetic change. Let us reiterate that drift may not be all-important simply because a trait *seems* to lack an adaptive significance, nor must it be necessarily considered inconsequential once the genes concerned are shown to have some influence on fitness. Drift is a mathematically defined certainty and selection is a biologically defined certainty. They play complementary roles in

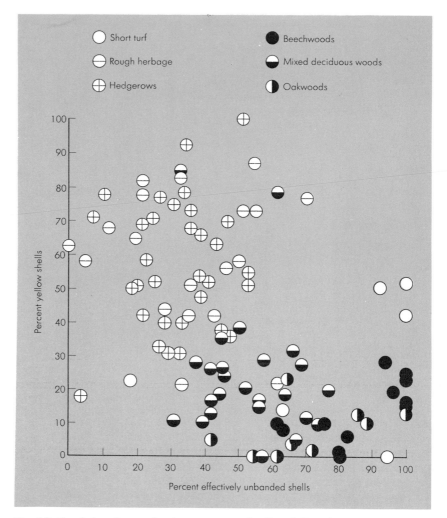

Fig. 6.3. Correlation diagram for the percentage of yellow shells, unbanded shells, and six kinds of distinct habitats. The proportions of the color and pattern types are positively associated with the character of each sort of habitat. Even the exceptions conform to the thesis of protective coloration (see text). For example, the exceptional mixed-deciduous-woods population was found in an atypical woods, which was carpeted with short green grass and had a few dead leaves. The high proportion of yellow forms is expected under such conditions. From the work of A. J. Cain and P. M. Sheppard in *Genetics*, **39** (1954), 99.

evolution, which are quantified by s and N in theory, but which are seldom separable in practice.

The role of small populations in evolution

Now that drift has been established by calculation and experimentation, but only in small populations, the question of its evolutionary importance might best be resolved by considering the role of small populations in evolution.

Most authors agree with Wright that drift in its classic aspect (i.e., chance fluctuations of gene frequencies in isolated populations that *remain* small) is probably of little consequence. Such populations will most likely be poorly adapted, because most loci will be homozygous, some being represented by selectively inferior alleles that drift to fixation. This loss of fitness, associated with homozygosity, is discussed in Chapter 7 from the standpoint of inbreeding depression, a common phenomenon accompanying self-fertilization or other sorts of close inbreeding in lines of normally outbred plants and animals. This concept applies here since consanguinity becomes more prevalent, regardless of the mating system, the smaller the breeding group (in fact, drift may be looked at as an inbreeding process and can be described in terms of the coefficient of inbreeding (see Chapter 3). Since deterioration becomes increasingly evident as more and more alleles become fixed with continued inbreeding and since variation will become minimal, selection having little store from which to effect an increase in the level of adaptedness, the result is "degeneration and extinction."

Now it is quite evident that extinction is the most probable fate of a population that has been reduced below a certain critical size or of one that is newly established by a few migrants settling outside of the normal range of the species. However, it should be noted that "inevitable" is a slight overstatement. Some highly inbred lines are known to persist for many generations. Likewise, a small isolate in nature may become specifically adapted to a particular habitat and maintain itself successfully for a prolonged period. A few are known that have done so. Nevertheless, such populations are constantly confronted with the possibility of changed conditions for which they cannot adjust, and they too are prone to extinction. But it may happen that a change would result in the extension of the colony's restricted niche and thus favor the expansion of the population. A great potential for evolutionary advance exists under these circumstances (see below).

Besides extinction or expansion, a small isolate has a third fate, that of merging with other populations. This is not of interest in the present context, however, since any differentiation established during isolation is quickly lost when free gene exchange is reestablished. The recurring theme in Mayr's magnificent treatise (Mayr, 1963) is that "gene flow and its consequences are essentially a retarding element as far as evolution is concerned."

It should also be recognized that inbreeding per se is not always detrimental to a species. There are many obviously well-adjusted plant forms that have adopted self-fertilization as their normal breeding system. Furthermore, it does not necessarily follow that inbreeding leads to homozygosity of all loci. The approach to complete fixation is a slow process, even when selection is not involved and inbreeding is extreme. That selection will be involved is most assured, and as stated earlier, selection favoring the heterozygous state will greatly dampen the effects of drift. For instance, in our laboratory, we purposely set out to make strains homozygous for different chromosomal inversions present in an experimental population of *Drosophila*. Scores of lines were maintained by sib-mating for more than twenty generations. But only one of several expected homozygous combinations ever became fixed; the great majority of lines remained polymorphic. The experimental population, which was derived from hybrids between two closely related species, has assumed a state of "obligate chromosomal heterozygosity" because of strong selection in favor of heterozygotes and has maintained a balanced chromosomal polymorphism for more than 10 years. There are numerous records of the retention of genetic variation in small isolated populations in nature.

A tremendous amount of evidence has accumulated (reviewed by Mayr, 1963) supporting the concept of geographic speciation. It is contended that isolated marginal populations, which originate from a few emigrants, are particularly subject to drastic genetic reconstruction and rapid evolutionary advance (explained in detail in Chapter 8). The opportunity for drift in such a situation is limited to the very beginning of the process (i.e., within the genetically improvished founding group). Since most surviving colonies are expected to increase in size, drift becomes a "one-shot" affair, resulting from zygotic randomness in the small sample of genetic variation carried by the migrants. This aspect of drift has been termed the *founder principle*. The point is that different founding colonies (or reduced populations) will possess, somewhat fortuitously, distinct (and depleted) genetic endowments. Once such a founding stock is established, selection will accumulate the fittest genotypes among those that are immediately formed. Thus, the path of evolution, although directed by selection, will be highly contingent on the initial genetic resources of the foundation colony; different paths will result from different base supplies.

Dobzhansky and Pavlovsky [see under Spiess (1962) in references to Chapter 5] have demonstrated the interaction of selection and the "founder" type of drift in experimental populations of *Drosophila pseudoobscura*. They showed that there is a greater variation in the outcomes of selection the smaller the starting populations are (Fig. 6.4). They first established a base stock consisting of F_2 individuals from hybrids between flies originating in Texas and flies from California. Those from Texas were homozygous for a particular gene arrangement in the third chromosome, denoted AR, while those from California were homozygous for a different sequence, symbolized PP. Thus, AR and PP chromosomes were expected to be in equal frequencies in the base stock. Next they established 20 experimental populations with random samples of individuals taken from the base stock. Ten of the populations were individually initiated with 20 "founders," while each of the other 10 was started with 4,000 flies. Within a couple of generations the "small" populations had increased to a size comparable to that of the "large" populations, and thenceforth all the populations were under similar selection regimes. The 20 populations were subsequently

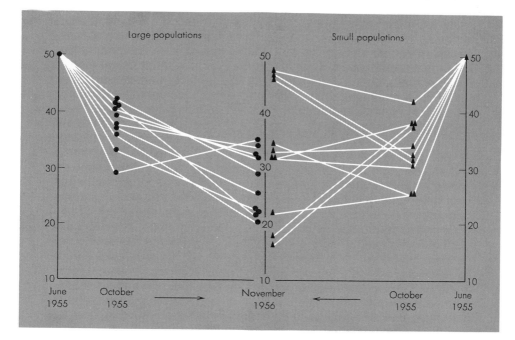

Fig. 6.4. The frequency (vertical scales equal percent) of PP chromosomes in 20 experimental populations of *Drosophila pseudoobscura* of mixed geographic origin (Texas and California). The ten replicate populations that were founded by small numbers of flies show far greater divergence after 21 generations than the continuously large populations. From the work of Th. Dobzhansky and O. Pavlovsky in *Evolution*, *11* (1957), 315.

analyzed twice: after 3 months and again at 17 months (about 21 generations of selection). As indicated in Fig. 6.4, the frequencies of PP chromosomes in those populations founded by few individuals ranged from 16 to 47 percent in the second analysis. The frequencies among "large" populations diverged less, ranging between 20 and 35 percent PP chromosomes. It was known that the frequencies of the PP and AR chromosomes in every population were adjusted by strong selection forces, which favored the heterozygotes PP/AR. However, the greater heterogeneity in results among the replicate "small" populations compared with the "large" ones can be taken as proof that the final outcomes of selection in the former were conditioned to some extent by genetic drift, which was obviously initiated in the founding generation. These results support the thesis that drift and selection may interact in producing evolutionary changes, even if the reduction in population size is just a temporary condition.

These workers have drawn upon the above facts to explain the divergence of certain populations in nature that conceivably have passed through a bottleneck in size, such as those founded by immigrants on small islands. An interesting case in point concerns the interpopulation variation of the meadow brown butterfly, *Maniola jurtina*,

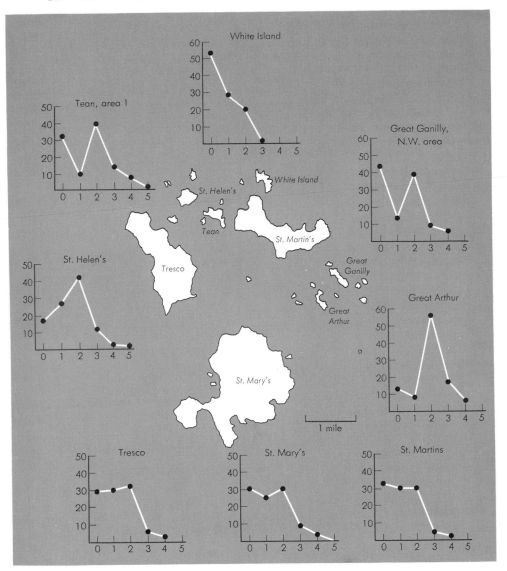

Fig. 6.5. The northeastern part of the Isles of Scilly. The three graphs at the bottom of the figure show the spot distribution (*vertical axis*, percent; *horizontal axis*, number of spots on hind wings) of female *Maniola jurtina* from the three larger islands and those encircling the upper portion of the figure represent populations from five smaller islands. Note the similarity of the "large-island" distribution and the diversity among the "small-island" distributions. After works by E. B. Ford, W. H. Dowdeswell, and K. G. McWhirter (see Ford, 1964).

on the Isles of Scilly off the southwest coast of England, reported by W. H. Dowdeswell and E. B. Ford (cf. Ford, 1964). Individuals of this species exhibit different numbers of minute spots (from 0 to 5) on the underside of the hind wings; populations can be characterized by the spot-distribution patterns that they exhibit. The spot distributions in the populations on the larger islands were observed to be nearly identical with one another, while those of the populations on certain smaller islands were remarkably different from each other and from the large-island populations (Fig. 6.5). With a few exceptions the characteristic spot distribution of the populations on all the islands has remained constant from year to year. It is unlikely that the similarity of the large-island populations has existed because of gene flow, since the interisland expanses are known to be effective barriers to migration and besides the small islands with their distinctive populations are interposed between the big ones.

Dobzhansky and Pavlovsky suggest that the small-island populations, although now large, were derived from a small number of immigrants; each small-island population started with a different genetic endowment, from which relatively stable but different gene pools were developed. Ford disagrees. He presents two bits of evidence which seem to directly refute drift in the given explanation. On one of the small islands (Tean), a large population assumed a radically new, and again stable, spot distribution without being reduced in size. The change was associated with an abrupt alteration in the vegetation of the island, brought on by the removal of a herd of cattle. Conversely, an effectively isolated population on one of the larger islands was reduced to near extinction during a year of drought, it subsequently increased with the recovery of the vegetation, but it retained its original spot distribution! Ford suggests that each population is adjusted to the conditions on each island; the butterfly populations on the larger islands are similar because they have been selected for the average of many conditions; each small-island population, on the contrary, is selected for different, rather uniform habitats. He claims that the strong selection pressures observed in nature (and expressed in breeding tests) are all that is needed to explain the *Maniola* variation pattern.

In conclusion, the mathematical theory of genetic drift has been well substantiated by experimentation. The debate on the evolutionary significance versus unimportance of genetic drift seems everlasting—at least until much more work and much less merely speculative discussion is done.

References

Ford, E. B., *Ecological Genetics*. New York: John Wiley & Sons, 1964.

Glass, Bentley, "Genetic Changes in Human Populations, Especially Those Due to Gene Flow and Genetic Drift," *Advan. Genet.*, *6* (1954), 95–139.

Glass, Bentley, M. S. Sacks, E. F. Jahn, and C. Hess, "Genetic Drift in a Religious Isolate: An Analysis of the Causes of Variation in Blood Group and Other Gene Frequencies in a Small Population," *Amer. Naturalist, 86* (1952), 145–160.

Lamotte, Maxime, "Polymorphism of Natural Populations of *Cepaea nemoralis*," *Cold Spring Harbor Symp. Quant. Biol., 24* (1959), 65–86.

Mayr, Ernst, *Animal Species and Evolution*. Cambridge, Mass.: Harvard University Press, 1963.

Mourant, A. E., "Human Blood Groups and Natural Selection," *Cold Spring Harbor Symp. Quant. Biol., 24* (1959), 57–63.

Stern, Curt, *Principles of Human Genetics,* 2nd ed. San Francisco: W. H. Freeman and Company, 1960.

Wright, Sewall, "On the Roles of Directed and Random Changes in Gene Frequency in the Genetics of Populations," *Evolution, 2* (1948), 279–295.

$\int e\mathcal{U}un$

Genetic Variation in Populations

To exist, a population must be adapted to its immediate environment. This means that most of the individuals in the population must possess genotypes and phenotypes which permit them to survive under the existing conditions. To persist, a population must be able to remain adapted by altering its genetic structure in response to the inevitable changes in the environment. This means it must produce and reproduce new genotypes and phenotypes adapted to the altered conditions.

The genetic flexibility, or plasticity, that enables a population to make an adaptive shift in phenotypes is maintained in different ways in different kinds of organisms. In microorganisms, with their enormous populations and very rapid reproductive rates, there seem to be enough newly arisen mutations at any given time to meet the challenge of novel situations through the rapid reproduction of those mutants which happen to be adapted to the new conditions. On the other hand, in multicellular, sexually reproducing organisms, whose populations are much smaller and whose reproductive rates are much lower, newly arisen mutations are seldom the basis for an adaptive response to changed conditions. Instead, genetic flexibility is achieved through the recombination of "stored" or concealed genetic variations already present in the population but unexpressed in the phenotypes of most individuals.

This chapter will be devoted primarily to a review of

137

some of the more important mechanisms known to affect the maintenance and utilization of genetic variation in sexually reproducing species.

Maintenance of genetic variation

Even though most populations are fairly well stabilized in their respective environments, the availability of preadapted genetic variants during times of environmental change is essential for continued evolution.

In diploid, sexually reproducing organisms, a surprisingly large amount of genetic variation can be maintained in the heterozygous condition without being expressed. This is possible because at many loci one allele carrying on normal activity is sufficient for the normal expression of the character.

On the other hand, the amount of concealed genetic variation that can be maintained in polyploids is greater than that for diploids, because one normal allele can mask the expression of several mutant alleles. But even though the amount of variation is greater, it may be less useful to the population because of greater difficulty in producing new phenotypes. For instance, in a tetraploid organism a recessive allele must be present in quadruplicate in order to be expressed. Thus, in having the potential to adapt to the more drastic environmental shifts, diploid organisms probably represent the best compromise between haploid organisms with little concealed variation and polyploid organisms with an excessively concealed but extensive amount of variation.

An understanding of the source and the maintenance of variation has held an important place in the history of the theory of evolution, and is still essential to our further understanding of evolution. Although evolution depends on the continued presence of genetic variation, one of its most important immediate consequences in a population is the inevitable production of ill-adapted individuals. The cost, in terms of reduced fitness of the population associated with the production of less than optimally fit individuals, is called the *genetic load*. As the noted British geneticist Haldane has reflected, "This is the price a population must pay for the privilege of evolution." Let us now consider how much variation actually exists in natural populations.

The detection of hidden genetic variation

The detection and quantification of hidden genetic variation is far from simple and is still an important problem in population genetics.

Most methods of detecting variation make use of systematic inbreeding to increase homozygosity which results in the expression of recessive alleles previously concealed in heterozygotes. Since many alleles are abnormal or harmful when homozygous, inbred strains usually show a lowered fitness or mean adaptive value ($\bar{W}$). This is especially true of species in which outcrossing is the rule. Falconer (1960, p. 249) lists several examples of inbreeding depression in populations of domestic animals. Generally, the more concealed variation a population has, the greater will be its inbreeding depression. In species that are normally inbreeders, in which inbreeding and selection have occurred for long periods of evolutionary time, there is little concealed variation and little, if any, reduction in adaptive value upon further inbreeding. But in outbreeding populations, the amount of variation concealed can be large and inbreeding depression severe. In principle, the amount of hidden genetic variation which is exposed upon inbreeding corresponds directly to the level of inbreeding obtained. For example, an F_2 generation produced by brother-sister mating will have an inbreeding coefficient (F value) of $\frac{1}{4}$. Consequently, on the average, one-fourth of the hidden genetic variation of the parental generation will be made homozygous and, therefore, will be expressed. In this case, there is approximately four times as much variation as was revealed by this amount of inbreeding.

To detect hidden genetic variation it is important to know what variants to measure. We could look for visible mutations and sometimes we do, but we know that these constitute a small proportion of new mutations and of the hidden recessives that have persisted in a population as well (for an excellent review of this aspect of hidden variation see Spencer, 1947). Detecting recessive lethals would give us a more objective idea of how much variation was concealed, but this too is a restricted group, and provides a limited measure of the total variation. Since most genes have some effect on the fitness of the organism, a more accurate estimate of total variation can be made by determining the reduction in the mean adaptive value (equivalent to average fitness) of the population with inbreeding.

In measuring changes in the mean adaptive value of a population, one would like to measure total fitness, but in practice this is hard to do. So inbreeding depression is usually expressed as the reduction of the mean value of some measurable component of total fitness, such as number of eggs laid, percent hatched, percent of individuals surviving to adulthood, etc. To illustrate, let us review the procedure for studying inbreeding depression in *Drosophila*, as measured in terms of viability from egg to adult. A sample of females, already fertilized, are collected from the wild and are placed individually in culture vials. The offspring of the females are systematically crossed to yield groups of individuals with different coefficients of inbreeding. For instance, the intermating of siblings, such as those produced in individual cultures of females from the wild, results in offspring with an F value of $\frac{1}{4}$. Matings between unrelated male and female progeny from different cultures produce essentially outbred offspring with an F value of 0. If offspring from one F_1 culture are outcrossed in separate matings to the offspring in each of two other F_1 cultures, then the offspring from each of these crosses will be cousins which, when interbred, will produce offspring with an F value of $\frac{1}{16}$. Groups with other inbreeding coefficients are produced by making appropriate crosses. How can you obtain double cousins whose offspring would have an *F* value of ⅛? Finally, the average survival (or mortality) value is determined for the outbred individuals (F = 0), and for each of the inbred groups having known *F* values. The average egg-to-adult survival values for various inbred groups, derived from several different natural populations of *Drosophila*, are presented in Table 7.1. These data, representa-

Table 7.1. The percentage of egg-to-adult survival with different degrees of inbreeding in *Drosophila* (in general there is a reduction of viability with an increase of inbreeding; the values in boldface are averages for several locality strains)

Species	$F = 0$	$F = \frac{1}{16}$	$F = \frac{1}{8}$	$F = \frac{1}{4}$
D. willistoni*	**84.3**	—	**76.0**	**63.5**
D. pseudoobscura†	**90.2**	**83.2**	—	**80.5**
D. pseudoobscura‡	88.7	—	71.4	73.6
D. pseudoobscura‡	88.0	—	81.9	78.2
D. arizonensis§	77.0	73.3	69.1	66.2
D. mojavensis§	62.4	46.1	48.4	44.1
D. ananassae†	90.8	75.0	—	65.5
D. ananassae†	85.2	79.3	—	63.4
D. ananassae†	70.9	71.0	—	67.2
D. novamexicana†	91.7	—	—	86.7

* From Ch. Malogolowkin-Cohen *et al.*, *Genetics, 50* (1964), 1299–311.
† From W. S. Stone *et al.*, *Genetics, 48* (1963), 1089–106.
‡ From Th. Dobzhansky *et al.*, *Genetics, 48* (1963), 361–73.
§ From L. E. Mettler (original data on laboratory populations).

tive of the many studies that have been made, show that there is a general decrease in survival with an increase in value of *F*.

   Similar kinds of data are obtained in man. In this case, the inbreeding coefficients are determined a posteriori from pedigrees, rather than by executing a series of prescribed matings. A typical example comes from marriage records and progeny mortality data for a rural French population filed by a church about a generation ago. The proportions of stillbirths and of infantile and juvenile deaths were listed for the progeny of consanguineous marriages and for offspring of reportedly unrelated parents. A portion of the data is as follows:

	Unrelated parents $F = 0$	*Cousin marriages* $F = \frac{1}{16}$
Stillbirths and infant deaths	$210/1,628 = 0.129$	$115/461 = 0.249$

Morton, Crow, and Muller (1956) have devised a method by which inbreeding data can be used to estimate the number of lethal equivalent mutants concealed in a population. Let us examine a simplified version of their approach, using the above human data. There is about a 12 percent excess of mortality (0.249 versus 0.129) among the progeny of cousin marriages over that of the outbred offspring. We have learned that $\frac{1}{16}$ of the genes that normally are heterozygous become homozygous in the progeny of cousins. Since there is a 12 percent increase in mortality when $\frac{1}{16}$ of the alleles are made homozygous, there should be approximately 16 times as much mortality if all persons were completely homozygous, i.e., if individuals could be formed by doubling each gamete. This means that there would be an increase in mortality of 200 percent ($0.12 \times 16 \approx 2$). It is estimated, therefore, that the average gamete in the sampled population carries two lethal equivalents. A "lethal equivalent" may be a single lethal mutant gene or a group of partially lethal mutant genes. This simply means that if an individual is homozygous for several genes, each of which reduces viability to some extent, their cumulative effect results in the death of the organism. Conversely, if the partially lethal genes were distributed in different individuals and made homozygous, they would cause, on the average, one death or its genetic equivalent in lowered fertility.

Since two gametes are required to produce a zygote, it is concluded from these data that the average person carries 4 lethal equivalents concealed in the heterozygous state. This is similar to *Drosophila,* in which the average number of lethal equivalents ranges between 0.4 and 4.0.

The accuracy of such an estimate and, accordingly, its biological significance, are contingent upon the validity of the assumptions underlying the model. One of the most critical specifications is that the genes that affect viability must act independently. In current terminology, we say that they act non-synergistically. If this is true, the effects of the alleles at separate loci will be multiplicative. (The probability of survival would be additive on a logarithmic scale, in which case log S, survival, is a linear function of F). Recent studies have shown that this is not always true, especially at high levels of inbreeding, as with the *Drosophila* data in Table 7.1. Very often, individuals with $F = \frac{1}{4}$ are nearly as viable as those with an inbreeding coefficient four times smaller, $\frac{1}{16}$. The problem of dealing with these and other discrepancies resulting from the assumption of independent action of loci is one of the most serious of current problems in population genetics (Dobzhansky *et al.,* 1963). But despite this and other shortcomings of the Morton-Crow-Muller model, we can use it to make rough estimates of the total genetic damage concealed in populations with respect to the fitness component measured.

In *Drosophila,* there is a more direct method of detecting some kinds of concealed variation, which involves making entire chromosomes homozygous through the use of special marker stocks. Unfortunately,

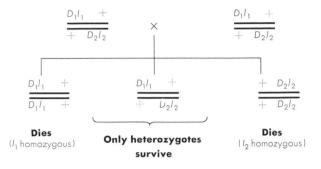

Fig. 7.1. A "balanced lethal system." Each mutant-labeled chromosome of the heterozygote D_1/D_2 carries recessive lethal factors, the expression of which is repressed by dominant wild-type alleles at corresponding loci on the alternate homologue. The simplest case involves at least two loci as depicted here. The lethal genes are linked in a state of complementarity ("trans" condition) by inversions that prevent crossing over. The homozygotes die due to the expression of homozygous recessive lethals. The strain D_1/D_2 therefore breeds true.

such stocks are available only in a few species of *Drosophila* and are not yet available in other genera. Typically, a tester stock is heterozyous for two different inversions (see Chapter 4), one in each member

Fig. 7.2. A generalized mating scheme used to make individual chromosomes homozygous, thereby revealing recessive mutants normally concealed in natural populations of *Drosophila*.

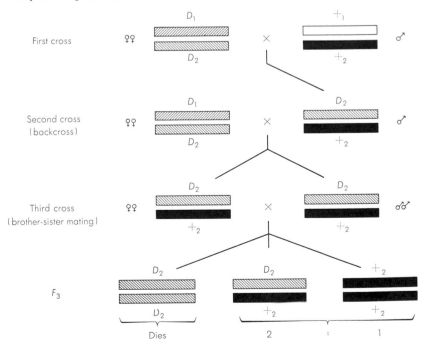

of a given pair of chromosomes. Each inversion is inseparably associated with a dominant phenotype and is lethal when homozygous. Two such markers commonly used in *D. melanogaster* are Curly and Plum. Used together, they constitute a balanced lethal system similar to that shown in Fig. 7.1. The only offspring that survive each generation are the *Cy/Pm* heterozygotes because *Cy/Cy* and *Pm/Pm* homozygotes die. Thus the stock is composed entirely of heterozygous individuals. To make a second chromosome homozygous, the scheme outlined in Fig. 7.2 is followed. A tester female is mated to a wild-type male. The offspring from this cross all will be heterozygous for one of the inversion chromosomes from the mother and one of the normal chromosomes (+) from the father. By crossing such an individual back to the tester stock, say *Cy/+* × *Cy/Pm*, a situation is established in which 3 of the 4 second chromosomes in the 2 individuals carry an inversion and a dominant marker. The offspring will be of 3 chromosome types, *Cy/Pm*, and *Cy/+* or *Pm/+*. The crucial point here is that each + chromosome now is a replicate of a single + chromosome from the previous generation. Thus by mating *Cy/+* × *Cy/+* this particular chromosome will become homozygous in ¼ of the zygotes. If the chromosome carries a lethal (or several partial lethals adding up to a lethal equivalent), these +/+ zygotes will die, leaving only *Cy/+* offspring. If the + chromosome is normal, the phenotypic ratio among the offspring of the third cross will be 1 wild type +/+ : 2 *Cy/+*, which phenotypically will be curly. Thus, in three generations we have extracted a single chromosome from a natural population, replicated it, maintained it intact (because the inversions prevented crossing over), and made it homozygous. Each sequence of matings, beginning with a single wild male, constitutes a test of one chromosome. Partial lethals (called subvitals, semilethals, detrimentals, etc.) result in a reduction of the wild-type class in varying degrees. Conventionally, if no offspring develop, the chromosome is considered a lethal. The tested chromosome is classified as a semilethal if a few but less than half of the expected number of +/+ offspring develop. Cultures in which more than half the expected one-third (or 16.7 percent) of +/+ flies develop have been variously referred to as quasinormal, subvital, and detrimental. Furthermore, if there are any mutations producing visible effects they will be expressed in the +/+ individuals.

In several species, similar tests can be made for the other chromosomes as well, and a composite picture of major chromosomes can be obtained. Results from this kind of study are somewhat higher than those obtained in the inbreeding studies.

As an example, data from experiments involving 113 homozygous and 90 heterozygous combinations of second chromosomes in *D. pseudoobscura*, sampled from a southeastern Arizona population, are presented graphically in Fig. 7.3. We see that the mean viability of the heterozygotes is 32.24 percent which is close to the expected proportion, ⅓. Note that most of the combinations tested actually were normal

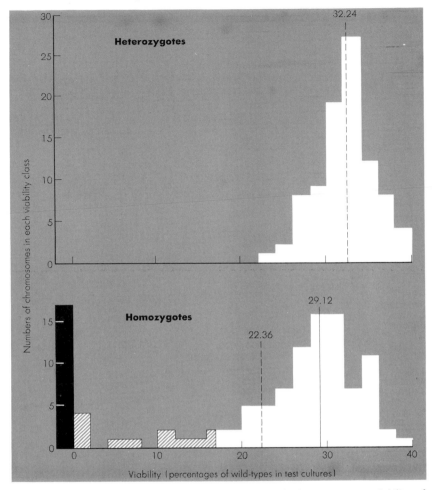

Fig. 7.3. The influence of **homozygosis** of different second chromosomes on viability of *Drosophila pseudoobscura* from Arizona. The numbers of second chromosomes giving various viabilities when homozygous and heterozygous are shown. The homozygotes are divided into three classes: lethals (black shaded), semilethals (diagonal hatching), and quasinormals (open). The mean values of the heterozygotes and all homozygotes (dashed lines) and of the quasi-normals alone (solid line) are given above the lines. Drawn from data presented by Th. Dob-zhansky and B. Spassky, *Genetics, 48* (1963), 1467-87.

with respect to viability, or very close to normal. However, some were distinctly subnormal. These probably result from a variety of causes, including chance homozygosis of allelic subvital recessives, and the occurrence of subvital genes that are partially dominant. Also, some of the heterozygous combinations tested were above normal. These are called "supervitals."

The fact that the average viability of the homozygotes is **22.36** per-cent, which is well below ⅓, indicates that there is an appreciable number of detrimental genes concealed in the outbred forms. Looking at it another way, nearly 24 percent (27/113) of the second chromosomes tested were either lethal or semilethal when homozygous. In a panmictic population, these chromosomes would be distributed at random. Each fly carries two second chromosomes. Since the frequency of chromosomes carrying a lethal or semilethal gene is 24 percent, the probability that both chromosomes in an individual are lethal or semilethal is $(0.24)^2$, or about 6 percent. Such individuals usually will not die, however, because most often the lethals or semilethals will not be allelic and will complement each other to produce normal offspring. Conversely, $(0.76)^2$ or 58 percent of the population will be entirely free of these grossly deleterious second chromosome mutations, and accordingly, 42 percent of the individuals carry at least one such chromosome. Similar tests can be made for the other two major pairs of autosomes. In the same Arizona population, it was determined that 25 percent of the third and 32 percent of the fourth chromosomes also carried lethals or semilethals. Consequently, the probability that an individual will be free of such genes on all its major autosomes is about $(0.76)^2 \times (0.75)^2 \times (0.68)^2 = 15\%$. There are no comparable data for the X chromosome.

Thus far we have considered only well-defined mutants with major effects on viability. A much larger portion of the hidden genetic variability consists of subvital chromosomes which contain mutant alleles with only slightly harmful effects. Subvitals are included in the quasinormal class, as outlined above. Their existence is readily deduced from the fact that the mean fitness of the quasinormals alone when homozygous, 29.12 percent, is significantly lower than that of the heterozygous controls, 32.24 percent (see Fig. 7.3). The fitness distribution is continuous, with no distinct differences setting off different viability classes. The proportion of chromosomes in a population, therefore, that one wishes to call subvital is arbitrary and must be established by convention. For this purpose, normality is often defined as those viability classes lying within two standard deviations (2σ) of the genetic variance of the heterozygous controls, that is, two standard deviations either above or below the mean. Viability classes falling more than 2σ above the mean are called supervitals. In most experiments, the subvitals (more than 2σ below the mean) make up a major portion of the quasinormals (Table 7.2). The lethals, semilethals, and subvitals make up the total load with respect to the viability component of total fitness. However, there are many other components of fitness, such as fertility (which is also shown in Table 7.2), fecundity, developmental rate, etc., that we have not discussed, which will contribute additionally to the hidden genetic variation.

In *Drosophila*, then, we can conclude that the average individual is, roughly, heterozygous for one visible mutation, two lethal equivalents, and four or more subvitals. Of course, these averages will differ in different populations, depending on the size, amount of inbreeding, severity of selection, ecological and geographical variables, and so

Table 7.2. The frequency of chromosomes in natural populations of *Drosophila* that reduce viability or fertility when homozygous*

Species	Chromo- somes	Lethal or semilethal	Subvital	Female sterility	Male sterility
D. prosaltans	second	32.6	33.4	9.2	11.0
	third	9.5	14.5	6.6	4.2
D. willistoni	second	28.4–41.2	57.5	40.5	64.8
	third	25.6–32.8	47.1	40.5	66.7
D. pseudoobscura	second	18.3–32.1	56.0–93.5	10.6	8.3
	third	16.4–25.0	29.5–86.7	13.6	10.5
	fourth	32.2	95.4	4.3	11.8
D. persimilis	second	25.5	84.4	18.3	13.2
	third	22.7	74.2	14.3	15.7
	fourth	28.1	98.3	18.3	8.4

* A range of values is given in those cases in which populations from different localities have been studied; subvitals are given as percent of quasinormals. Compiled from the work of Dobzhansky and collaborators.

forth. But in general, it can be seen that the amount of genetic variation and, consequently, the potential for evolutionary change, is enormous.

The adaptive norm and the concept of genetic load

Let us now examine the costs to a population of maintaining this variation. Ideally, each individual in the population should have the genotype and phenotype best suited for the immediate environment. But we have come to realize that, for a variety of reasons (see section on homeostasis and canalization below), a great variety of genotypes will produce a rather limited number of fairly similar phenotypes. This is inherent in the concept of concealed genetic variation. It follows that a variety of different genotypes can be equally well adapted. (It is this latter fact which frustrates our efforts to determine the exact amount of variation in a population, because it is so difficult to detect the presence of this kind of variation). Consequently, a majority of the genotypes in a population will produce well-adapted and "normal" individuals. It is this group of normal individuals which we call the adaptive norm. The concept is represented in Fig. 7.4 by the dotted curve, with the region *IJ* indicating a plateau under which there are a number of genotypes and phenotypes that are equally well fit, or nearly so. The concept of the adaptive norm rejects the typological species concept (Chapter 2), which suggests that there is a single "best genotype" and "best phenotype" producing a population composed largely of virtually identical "typical" individuals. To

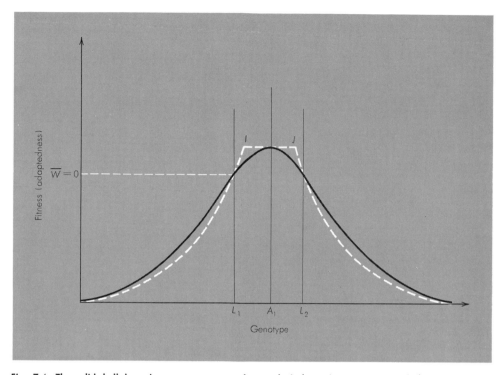

Fig. 7.4. The solid bell-shaped curve represents the typological species concept in which there is a "best" genotype (and best phenotype) at A_1. L_1 and L_2 represent environmentally imposed limits of survival under normalizing selection. The dashed line represents the concept of the *adaptive norm*, in which there are many genotypes producing similar phenotypes having equal fitness.

quantify the concept of adaptive norm, Wallace and Madden (1953) have defined it to include those individuals that fall within two standard deviations of the mean fitness of a representative sample of heterozygous individuals. Even though there is no "best genotype" or "best phenotype," for short-term adaptedness it is still advantageous for every individual to be well adapted to his immediate environment. But, as we have mentioned, the environment will inevitably change, so that for purposes of long-term survival the population must constantly produce some individuals which are not well suited to the present environment, but which by chance may be preadapted to some future environment. In this way, a population can easily shift so as to remain adapted to its environment as conditions change. As mentioned previously, variation which gives rise to the ill-adapted phenotypes is the genetic load.

The concept of genetic load has undergone considerable development since Haldane first called attention to its existence, and since Muller first used the term in the late 1940's. It is not too difficult to accept that the large amounts of genetic variation in a population must be maintained for continued adaptation, and that this will inevitably reduce the immediate, over-all fitness of the population by producing ill-adapted individuals. Genetic load, in its broadest sense, simply refers to any reduction in population fitness, actual or potential, resulting from the presence of genetic variation. However, a number of factors complicate the subject, among them the different kinds of mutant genes (lethals, semilethals, detrimentals, etc.) which contribute to the load, and different modes of selection which exact the cost.

Although grasping the concept of genetic loads is not overly difficult, the measurement of their magnitude, the determination of their actual effects on populations, and even defining them precisely have proved to be very hard problems to solve. Indeed, there has been no general agreement, as yet, on how even to approach them. Crow (1958) has suggested that genetic load be defined "as the proportion by which the population fitness [or whatever other trait is being considered] is decreased in comparison with an optimum genotype." This definition has been criticized on the grounds that it is hard to identify an optimum genotype, that the choice of an optimum is quite arbitrary, and that the magnitude of the load could, theoretically, change drastically simply by introducing a single new beneficial mutation into the population. The extent to which such criticisms are valid has not been settled. Nevertheless, this definition has served as a focal point for all recent attempts to deal with genetic loads more quantitatively.

Dobzhansky (1964) has defined the genetic load in terms of the adaptive norm of the population. The definition is based on the average fitness of the heterozygous individuals in the population. Those individuals with fitnesses falling more than two standard deviations below the mean of the heterozygotes are the bearers of the genetic load.

Let us now examine genetic loads in more detail.

Mutation and mutational loads

Basically, there are three kinds of loads: mutational, balanced, and substitutional (or transient). Each of these will add to the total genetic load of a population, because some loci will contribute to the mutational load, others to the balanced load, and still others to the transient load. The mutational load is that fraction of the total load which results from mutant alleles which are present in very low frequencies because they are selected against (normalizing selection), but which persist because of recurrent mutation. This is what has usually been considered a typical locus. Let us consider more fully the effects at these loci of mutation and selection acting simultaneously on a population.

In Chapter 4 (just preceding Eq. 4.5) we saw that, by considering the effects of mutation alone, $\Delta p = -p_t u + (1 - p_t) v$. Conversely, $\Delta q = p_t u - q_t v$. Considering the effects of selection against a recessive genotype (Eq. 5.5) we saw that

$$\Delta q = -\frac{sq^2(1-q)}{1-sq^2}$$

As a result, we have mutation tending to increase the frequency of q on one hand, and selection tending to decrease it on the other. When these two opposing forces come to equilibrium, as they must,

$$pu - qv = +\frac{sq^2(1-q)}{1-sq^2}$$

At this point it is customary to assume that in any real population the frequency of an allele responsible for a detrimental phentoype will be small, in this case q. Therefore, we can drop out the term qv (approximately zero) and consider the term $1 - sq^2$ to be approximately unity. Thus:

$$pu = sq^2(1-q)$$
$$(1-q)u = sq^2(1-q)$$
$$u = sq^2 \quad \text{(approximately)} \qquad \text{(Eq. 7.1)}$$

$$q^2 = \frac{u}{s}$$

$$q = \sqrt{\frac{u}{s}} \quad \text{(approximately)} \qquad \text{(Eq. 7.2)}$$

At such a locus the proportion of affected individuals (q^2) times the extent to which they are ill-adapted (s) constitutes the expressed mutational load (sq^2 in Eq. 7.1).

Note that, for any given s, this is equal to the mutation rate u.

Equation 7.2 actually applies only when an allele A is completely dominant and selection is against the homozygous recessive, aa. When selection is against the dominant allele,

$$\Delta q = \frac{sq^2(1-q)}{1-s(1-q^2)}$$

and, at equilibrium,

$$-(pu - qv) = \frac{sq^2(1-q)}{1-s(1-q^2)} \qquad \text{(Eq. 5.7)}$$

This time it is pu that is negligible; the denominator, $1 - s(1-q^2)$, is again considered unity. Solving for $2pq$ we have:

$$qv = sq^2(1-q)$$

$$\frac{q}{q^2(1-q)} = \frac{s}{v}$$

$$q(1 - q) = \frac{v}{s}$$

$$2pq = \frac{2v}{s} \quad \text{(approximately)} \qquad \text{(Eq. 7.3)}$$

When there is no dominance, and allele a is selected against,

$$\Delta q = \frac{\frac{1}{2}sq(1 - q)}{1 - sq} \quad \text{(equation from Falconer, 1960, p. 30)}$$

and at equilibrium,

$$q = \frac{2u}{s} \quad \text{(approximately)} \qquad \text{(Eq. 7.4)}$$

To calculate the mutational load for varying degrees of dominance (Crow, 1958) consider a locus where the genotypes AA, AA', and $A'A'$ have relative fitnesses of 1, $1 - hs$, and $1 - s$ respectively. The degree of dominance of the A' allele is represented by h. In this case, the simplifying assumption is that since the frequency of $A'A'$ homozygotes will be very small we can ignore their contribution to the mutational load, i.e., selection against A' will be in the heterozygotes. At equilibrium the frequency of heterozygotes will be reduced by an amount hs every generation, but these will be replaced by mutation from $A \rightarrow A'$ in the homozygous AA class. Since each individual has two A alleles this rate will be twice the rate per allele, or $2u$. Thus $(2pq)(hs) = 2u$. Again the mutational load is seen to be dependent on mutation rate alone, but except where h is nearly zero it will be equal to twice the mutation rate. When $h = 0$, Eq. 7.1 applies. To obtain a total mutational load, we must sum the loads for individual loci.

It has not yet been possible to determine in any actual population what proportion of the total load is mutational, although this is a crucial issue at present. Some authors feel that it constitutes the major portion, whereas others feel it makes up a minor part, with the balanced load (see next section) constituting the major fraction. The extent to which one or the other contributes to the total load depends both upon the proportion of loci that are maintained by balancing selection instead of normalizing selection (i.e., the extent to which the situation we have just described is really typical) and upon the magnitude of the load per locus for each type of selection. We should also note that a change in the environment could shift the mode of selection in such a way that a locus would be shifted from contributing to the balanced load to adding to the mutational load or vice versa. This type of shift can be clearly illustrated in the case of sickle cell anemia. In those regions of Africa having a high incidence of subtertian malaria, there appears to be strong balancing selection which maintains the sickle cell allele at a high frequency. This is because the heterozygotes are resistant to the malaria and are only slightly anemic. Individuals of one homozygous genotype, Sk/Sk, are highly susceptible to malaria, though not at all anemic, and individuals of the other homozygous genotype, *sk/sk*, are severely anemic. In other parts of the world, such as in North America, that have no malaria problem, the *sk* allele occurs in low frequency. These regions continue to have strong selection against the essentially lethel *sk/sk* genotype, but here selection is normalizing, because the *Sk/Sk* individuals are superior to the slightly anemic heterozygotes. In these areas q is much lower, and the locus contributes to the mutational load.

Balancing selection and balanced loads

In studying natural populations, one is struck by the multitude of instances, such as in sickle cell anemia in the ABO blood groups, or in many enzyme systems (Fig. 7.5), in which two or more alleles at a locus are maintained at a high frequency. The various genotypes and their phenotypes also occur with relatively high frequencies. Such polymorphic loci differ from our "typical" locus not in the number of different phenotypes which are produced but in their frequency. These loci are usually maintained by some form of "balancing" selection, which involves opposing selection forces. We have previously seen, for example, that a balanced equilibrium results when an allele selected against in the homozygous state is retained because of the superiority of heterozygotes (overdominance). Other balanced states may occur when:

1. an allele is favored at one developmental stage and is selected against at another. (A special case of this can be seen in "meiotic drive" in which distortions of meiosis produce gametes with one type predominating.)
2. an allele is favored in one sex and selected against in another.

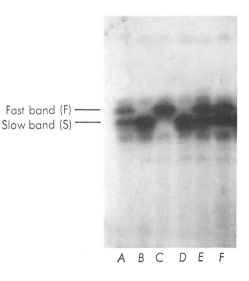

Fig. 7.5. In recent years, with the development of new biochemical techniques, many enzyme polymorphisms have been discovered. An example is esterase-6 in *Drosophila*, which is under the control of the esterase-6 locus, on the third chromosome. In all populations, fast and slow variants of the enzyme have been found with heterozygotes having both. In the photograph above, the enzymes have been separated by starch gel electrophoresis in *D. ananassae* and made visible by activating them with appropriate substrates. The esterase-6 bands are those which stain most darkly. The lighter bands are other kinds of esterases. Individuals *A, E,* and *F* are heterozygous (*FS*); *B* and *D* are slow (*SS*); and *C* is fast (*FF*). Courtesy of F. M. Johnson.

Fast band (F) ——
Slow band (S) ——

A B C D E F

3. an allele is favored at one time (i.e., in one season) and is selected against at another.
4. different genotypes are able to exploit different environmental situations within a habitat, and thereby reduce competition and elimination through selection. This is called the Ludwig effect.
5. different alleles are favored in subpopulations exposed to different environments characteristic of adjacent habitats.
6. an allele is favored when it is rare and selected against when it is common.

Each of these, with the possible exception of No. 4, will result in genetic loads which we call balanced loads.

Let us consider further the case in which the polymorphism is maintained by overdominance. We could cite numerous examples in which the heterozygote can be shown to be superior in fitness to either homozygote. Why this should be true is certainly far from clear in most instances (see James L. Brewbaker, *Agricultural Genetics*, in this series). However, one way in which it could occur can be seen if we consider for a moment the mechanisms of gene action. A gene, in conjunction with ribosomes and such molecules as messenger RNA, transfer RNA, and various related enzymes, exercises its control in an organism by specifying the amino acid sequence in specific functional proteins such as enzymes, antibodies, etc. If there are two different alleles possible at a locus, the heterozygote is potentially more versatile than either homozygote in that it might produce two different polypeptides instead of only one. This could be advantageous in several ways. If both alleles produced proteins that were useful at all times, a carrier of both would be at a selective advantage. Or perhaps the two alleles are useful at different times during the life cycle, or are useful under different environmental conditions or different intracellular conditions. In this connection, it is interesting to note that in recent years a number of investigators have discovered that what were formerly believed to be single pure enzymes often are actually groups of closely related enzymes. In a single individual these groups are very similar and catalyze the same reaction, but are not identical in structure or in requirements for optimum activity. These similar enzymes have been called *isoenzymes* or *isozymes*, and each complex may contain up to a dozen or more. Enzyme activities are known to be greatly influenced by relatively small changes in such factors as temperature, pH, and concentration of various ions. It is presumed that two or more isozymes can serve the needs of a cell, or of an organism, more efficiently and over a wider range of environmental conditions than could a single enzyme. Possibly these are examples of balanced polymorphism. The isozyme examples strengthen the supposition that two similar but nonidentical enzymes could stabilize an individual over a wider range of conditions than would be possible with either enzyme alone. Also, it is becoming more apparent that interactions between genes at different loci are especially important, and a heterozygote could interact in more ways than could a homozygote. However, the fact of the matter is that, for single-locus heterosis, our knowledge is scant and our explanations speculative. This lack of knowledge is further complicated by the fact that heterosis very frequently involves alleles that are considered to be recessive and which are lethal when homozygous. We surmise, though perhaps erroneously, that the majority of these are inactive, and indeed that their recessiveness is due to their inactivity. How, then, can such genes result in heterosis?

A more complicated kind of heterosis has been studied extensively in *Drosophila*, especially by Dobzhansky and his co-workers. Instead of affecting single genes, it involves gene complexes that are contained within and protected by chromosomal inversions.

In natural populations, it is often true that most of the individuals caught in the wild are heterozygous for different inversions. In well studied cases, such as those of *D. pseudoobscura* by Dobzhansky and others, it has been found that the inversions do not occur haphazardly, with each one different, but that there are a number of distinct inversions widely distributed and occurring with high frequencies in many populations throughout the species range. Some of the more common of these, many of them named after the locality in which they were first collected, are Standard, Arrowhead, Chiricahua, Santa Cruz, Treeline, Olympic, Estes Park, and Pikes Peak. From specimens collected from many areas in northern Mexico and the southwestern United States it becomes obvious that, although there are a number of inversions in most populations sampled, one or two are predominant. Furthermore, Standard, which is predominant along the coast (see Fig. 7.6), is still common in inland California and becomes rare in the rest of the species range. Chiricahua is predominant on the Mexican plateau, common in California, and rare in the rest of southwestern United States. Arrowhead is predominant in Utah, Colorado, Arizona, and New Mexico, whereas Pikes Peak is predominant on the eastern slope of the Rockies and western Texas.

It has also been shown that in mountainous regions there is a distinct change in chromosome frequency with altitude (Fig. 7.7). For instance, in the Sierra Nevada of California, the three major inversions change as follows. The frequency of Standard decreases progressively from 46 percent at an elevation of 850 feet to 10 percent at 10,000 feet. Arrowhead changes from 25 percent at 850 feet to 50 percent at 10,000 feet and Chiricahua fluctuates somewhat in going from 16 percent at 850 feet, to 20 percent at 10,000 feet.

Another series of experiments has shown that, within particular populations, there is seasonal variation in the frequency of some inversions. In a population from Mt. San Jacinto, California, the frequency of the Standard chromosome arrangement is high in the early spring, about 53 percent, drops to a low of about 27 percent in June, and slowly climbs back up to 50 percent by October. The frequency

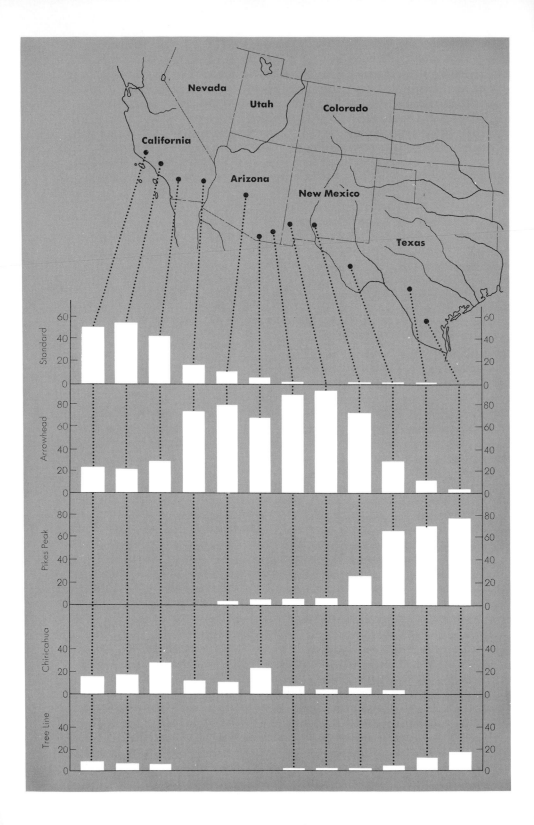

Fig. 7.6. Percentage frequency of different inversion of chromosome III in a series of populations of *Drosophila pseudoobscura* along a transect from southern California to southern Texas. After Th. Dobzhansky and C. Epling, *Carnegie Inst. Washington Publ. No. 554* (1944).

of the Chiricahua chromosome does just the reverse. It starts out in March, when the population first increases from its overwintering low, with a frequency of about 23 percent, climbs to a high of 40 percent in June, and recedes again gradually to 20 percent by October (Fig. 7.8).

These observations all indicate that the different gene arrangements have different adaptive values under different conditions. This has been confirmed by a number of laboratory experiments in which different chromosomal combinations were exposed to different environmental conditions. Most of the experiments involved differences in relative humidity or temperature, but some involved other factors as well, such as limited diet. At 100 percent relative humidity, the Chiricahua

Fig. 7.7. Frequencies of two inversion types (St and Ar) of chromosome III in populations of *Drosophila pseudoobscura* at different elevations in the Yosemite region of the Sierra Nevada. The localities sampled are shown on the profile map. The percentage frequencies of St and Ar chromosomes in populations at each locality are given above the profile. From V. Grant, *The Origin of Adaptations.* New York: Columbia Univ. Press, 1963. Redrawn from Th. Dobzhansky, *Genetics, 33* (1948), 158–176.

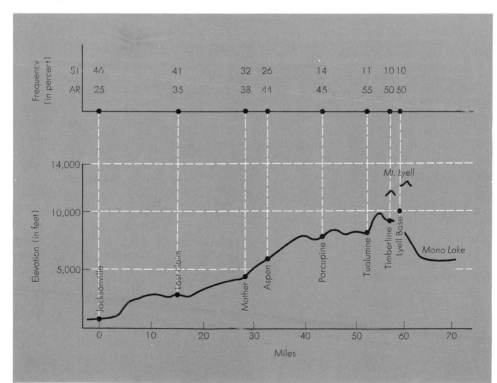

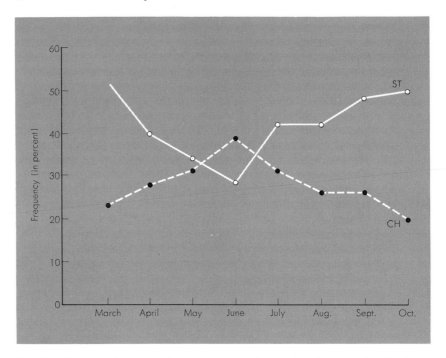

Fig. 7.8. Changes in frequency of chromosomes carrying two inversion types in natural populations of *Drosophila pseudoobscura* in the San Jacinto Mountains during the advance of the season from March to October. From V. Grant, *ibid.* (see Fig. 7.7).

sequence is more viable (91 percent) than either Arrowhead (74 percent) or Standard (84 percent). At 92 percent relative humidity, Standard is the most viable (94 percent versus 81 percent and 85 percent for Chiricahua and Arrowhead) and at near 0 relative humidity Arrowhead is most viable (74 percent versus 54 percent and 59 percent for Chiricahua and Standard). At temperatures of 28° to 30°C., St/St is most viable, and St/Ch is so at temperatures of 0° to 4°C. Although many tests show the heterozygotes to be more viable than either homozygous type, this is not invariably true, as these figures indicate. The observations correspond well with a large number of population cage experiments both in *D. pseudoobscura* and in a number of other species. Under these conditions, there is competition for survival at every stage in the life cycle and the relative survival values (including reproduction) of different inversions can be tested. When several inversions are introduced into a population cage under given conditions, an equilibrium is reached with each inversion achieving a characteristic frequency. Significantly, the equilibrium is always established near the same value, regardless of the initial inversion frequencies. However, by changing the cage conditions, the equilibrium point can be altered. In fact, the relative fitness of two inversions can be altered by the addition of a third inversion in the initial population.

Recall from Chapter 4 that the genes within an inversion will be inherited as a block if most of the individuals in the population are inversion heterozygotes, and they will be acted on by natural selection as a group, rather than as individual genes. We have seen that there is abundant evidence indicating that the inversions in *D. pseudoobscura* have high adaptive value under certain conditions and lower adaptive value under others. The adaptive value of a particular inversion differs from one geographical locality to another. For instance, four different cage populations, started with 20 percent *St* and 80 percent *Ch*, all came to equilibrium with the frequency of *St* being 80 percent (Fig. 7.9). It can also be seen that the equilibrium was established at the same rate in each population. All four populations were begun with chromosomes of the same geographical origin, Pinion Flats, California. The equilibrium indicates that *St/Ch* heterozygotes are superior to both homozygous combinations. These results are in marked contrast to those shown in Fig. 7.10 where the *St* chromosome was from Pinion Flats, but the *Ch* chromosomes were from six different Mexican populations. In this case there was no uniformity. By the time the experiment was terminated, four populations had gone to fixation for the *St* chromosome, but at different rates, and two appeared to have come to

Fig. 7.9. Four population cages of *Drosophila pseudoobscura* started with 20 percent Standard (*St*) and 80 percent Chiricahua (*Ch*) chromosomes from the same geographical area, Pinion Flats, California. Note that they all change uniformly until an equilibrium is reached. From Th. Dobzhansky, *Proceedings of 9th International Congress of Genetics* (1945), 435–49.

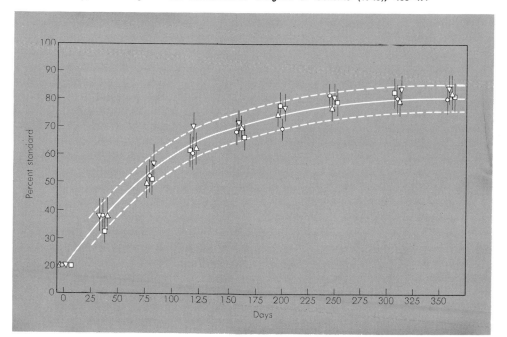

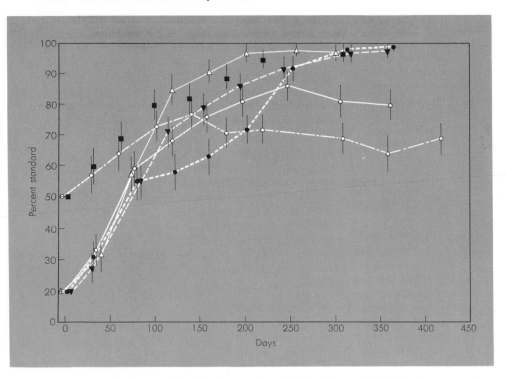

Fig. 7.10. An experiment like the one shown in Fig. 7.8, except that the chromosomes have come from different geographical areas. In this case the results are not uniform, and in these cages the Ch chromosome is lost entirely. The genetic composition of the same inverted region obviously differs from place to place. From Th. Dobzhansky, *Proceedings of 9th International Congress of Genetics* (1954), 435–49.

equilibrium, but at different levels. Obviously, the *Ch* inversion is genetically different in different geographical regions.

A reasonable conclusion to draw from all these observations, as Dobzhansky has done (1951), is that the inversions protect gene complexes from crossing over, so that an integrated group of genes can be built up over an evolutionary period of time and subjected to natural selection as a group (as supergenes). It would appear, in *D. pseudoobscura*, at least, that different complexes have become adapted to different facets of the local environment. An individual heterozygous for two such complexes apparently is better buffered against environmental fluctuations because it will have two integrated and internally balanced gene complexes, each of which is able to cope with a different, or at least partly different, set of environmental conditions. This gives rise to an inversion polymorphism which is adaptive. In other words, the polymorphism itself is adaptive because it enables a population to better survive the uncertainties of the environment than would be possible on the basis of a single genotype or of a single coadapted gene complex.

Limits to number of overdominant loci

The maintenance of polymorphisms through overdominance is one type of balancing selection that keeps two or more alleles present in a population with high frequencies. The cost of maintaining polymorphism by overdominance, in terms of genetic load, can be very high. At equilibrium,

$$\hat{q} = \frac{s_1}{s_1 + s_2} \qquad \text{(Eq. 5.10)}$$

Under these conditions, a number of less well adapted homozygous individuals will always be produced from segregation in the heterozygotes. This constitutes a portion of the balanced load and possibly a major portion. Crow (1958) has called this the "segregational load" because it arises from segregation in the heterozygote. If the relative fitnesses of AA, AA', and $A'A'$ are $1 - s_1$, 1, and $1 - s_2$ this load is equal to $s_1 p^2 + s_2 q^2$, the extent to which both homozygous classes are reduced in fitness. Using the $\hat{q}$ formula above and knowing that

$$\hat{p} = \frac{s_2}{s_1 + s_2}$$

we can substitute for p^2 and q^2 as follows:

$$
\begin{aligned}
\text{Segregational load} &= s_1 \frac{(s_2)^2}{(s_1 + s_2)^2} + s_2 \frac{(s_1)^2}{(s_1 + s_2)^2} \\
&= \frac{s_1 (s_2)^2 + s_2 (s_1)^2}{(s_1 + s_2)^2} \\
&= \frac{s_1 s_2 (s_1 + s_2)}{(s_1 + s_2)^2} \\
&= \frac{s_1 s_2}{s_1 + s_2}
\end{aligned}
$$

At a locus where both homozygotes are lethal, the load will be very heavy. Indeed, 50 percent of the offspring fail to survive. This is an extreme example rarely encountered in nature, but it has raised the very pertinent question as to how many overdominant loci a population can maintain without incurring an unbearable load. For instance, if only heterozygotes survive, the fraction of survivors in each generation due to this locus alone is reduced to $\frac{1}{2}$. Suppose that there were 10 such loci in the population. Then the probability that an individual is independently heterozygous at all 10 loci, and capable of surviving, is $(\frac{1}{2})^{10}$ or $1/1,024$. This means that only one zygote out of 1,024 will escape genetic lethality. One doubts that there are many species with the reproductive potential to survive selective pressures such as these extremes. Of course, the assumption that the loci are inherited independently is gratuitous. They could all be linked, or some of them could be linked, in which case the load would be reduced because of the tendency for lethal combinations, i.e., homozygotes, to be concentrated in fewer individuals with a corresponding increase in the probability that a given zygote would be free of lethal combinations. But even so, the number of overdominant loci that could be tolerated would be severely limited.

The load would be considerably less severe if the homozygotes were not lethal. Suppose that each homozygote, instead of dying, reproduces 98 percent as efficiently ($s_1 = s_2 = 0.02$) as the heterozygote, in terms of number of offspring reaching reproductive maturity. Then, by the same reasoning as above, the probability that a zygote escapes genetic death at this locus is 396/400 (at equilibrium $q = 0.5$). The zygotic proportions at the beginning of each generation would be $\frac{1}{4}AA : \frac{1}{2}Aa : \frac{1}{4}aa$ or 100 : 200 : 100 to take convenient numbers. But the proportions at reproductive maturity would be 98 : 200 : 98 from which the probability that a given individual will survive is calculated (0.99). If there are 10 such independent loci, the probability of survival would be $(0.99)^{10} = 0.904$, for 100 loci $(0.99)^{100} = 0.363$, or for 1,000 loci $(0.99)^{1,000} = 0.000043$. Given the above calculations, one can reasonably conclude that not many more than 100 polymorphisms can be maintained by overdominance in most organisms, even if our estimates of lethality are somewhat too high. Most geneticists agree that even a relatively simple organism like *Drosophila* has 10,000 or more loci. If this is true, then far fewer than 10 percent of the loci can be overdominant even if the homozygotes are only slightly at a disadvantage. The logic of this assertion seems impeccable. However, a lively discussion has arisen concerning this point within the last few years. It was previously asserted by Wallace (1958) that 50 percent or more of all loci in *Drosophila* were heterozygous. These results were derived from a radiation experiment and their general applicability was regarded with some suspicion. More recently, however, the landmark studies of Johnson and co-workers (1966), and Hubby and Lewontin (1966), represent the first direct attempts to determine what proportion of loci of a species in the wild state are polymorphic. Using recently developed starch gel electrophoresis methods for studying enzyme variability, Johnson's group tested *D. ananassae,* and Lewontin and Hubby, *D. pseudoobscura.* Lewontin and Hubby observed the variation in 21 enzymes for which assay methods were available. It has been repeatedly demonstrated that the production of enzymes is under direct genetic control, and on the basis of the extensively supported and widely accepted one gene–one enzyme (or better, one cistron–one polypeptide) theory (see Philip E. Hartman and Sigmund R. Suskind, *Gene Action,* in this series), these would represent 21 loci. Hubby and Lewontin discovered, over the whole species range, that 9 of the 21 loci (or 39 percent) were polymorphic. The number of loci that were polymorphic in any given locality was 30 percent, on the average.

Based on these data—which for various reasons, such as small sample size, underestimate the number of polymorphic loci—a conservative estimate of the number of polymorphic loci in *Drosophila* is considered to be 2,000. These results were quite consistent with those in *ananassae*. If we accept the observations and hold to our previous conclusion that not many more than 100 loci can be maintained by overdominance, then it must be true that most of these polymorphic loci are maintained by other mechanisms. This is essentially what Hubby and Lewontin have suggested. However, several others (Sved, Reed, and Bodmer, 1967; King, 1967) have challenged the original assumption that each locus acts independently in reducing fitness. This would mean that, in our previous example, $(0.99)^n$ does not satisfactorily represent the probability of survival. Although their arguments are based on advanced models, the reader with a moderate background in mathematics would find these papers interesting. Assuming independence, being homozygous for two loci, each with a fitness of 0.99, would render the carrier twice as likely to die (actually the value is 0.0199 rather than 0.02, because of the probability of 0.0001 that both independent lethal effects would occur simultaneously) than if the carrier were homozygous for only one of the loci. Under the more reasonable assumption of functional interactions between loci it is possible to propose a situation in which each additional homozygous locus would be progressively less deleterious. A level is soon reached after which additional homozygous loci fail to decrease fitness any farther.

In this regard, another important development in the area of genetic loads, which helps explain some of the seeming contradictions between the amount of heterozygosity and the size of the load, has been reported by Mukai (1968). He finds that for at least a certain class of genes called polygenes there is an optimum number of heterozygous loci. A polygene is defined as a gene which has a relatively small effect on fitness, i.e., it is not lethal or semilethal. The experimental procedure was as follows. A single second chromosome was extracted from a natural population of *D. melanogaster* and expanded into a number of lines, using the mating scheme shown in Fig. 7.11. The accumulation of polygene mutations was followed in each line by periodically testing the viability of the second chromosome. This was done by mating $Pm/+_i$ to Cy/Pm and intercrossing $Cy/+_i$ offspring as shown in the lower half of Fig. 7.11. To test for heterozygous viability, $Cy/+_i$ was mated to $+_0/+_0$ from a tester line in which $+_0$ was known to have normal viability. Over a period of 25 generations the homozygous viability declined steadily as polygenic mutations accumulated at an average rate of 0.1411 per second chromosome per generation. (This is much higher than the rate for lethal mutations, which was 0.006). On the other hand, heterozygous viability varied. If the test was made in such a way that the rest of the chromosomes (genetic background) are homozygous, then the heterozygous viability of the tested chromosome was higher than normal, indicating that the mutant loci were

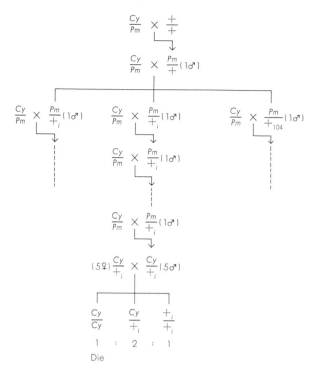

Fig. 7.11. Mating scheme for establishing and perpetuating numerous independent chromosome lines for the second chromosome in *Drosophila melanogaster*. All lines are replicates of a single original second chromosome extracted from a wild population. From T. Mukai, *Genetics,* **50** (1964), 1–19.

overdominant. If the genetic background was already heterozygous the heterozygous viability of the test chromosome was at best normal, though it usually is below normal. Thus the effect of the new mutations on fitness depends on how much heterozygosity is already present. If heterozygous viability is plotted against homozygous viability, when the genetic background is homozygous, the results shown in Fig. 7.12 are obtained. In Fig. 7.12a the mutant genes are all in coupling phase (all on the same chromosome) in the heterozygote. It can be seen that as the number of mutants increases, resulting in decreased homozygous viability, the heterozygous viability increases to a maximum and then decreases again. The point of optimum heterozygosity corresponds to about 9 loci. Quite surprisingly, when the heterozygous viability of polygenic mutations was tested in "repulsion phase" (mutants present on both chromosomes) it declined right along with the homozygous viability. This is shown in Fig. 7.12b where $+_i/+_j$ represents chromosomes from two different experimental lines. In this case the viability of $+_i/+_j$ is plotted against the standardized average of the viabilities of the two kinds of homozygote, $+_i/+_i$ and $+_j/+_j$. Whatever the cause for this odd phenomenon, it may provide an explanation for conflicting observations of Wallace (1958) and Greenberg and Crow

Fig. 7.12. The "coupling-repulsion" effect and optimum heterozygosity for polygenes effecting viability in *D. melanogaster*. The two graphs (*a* and *b*) show the relative viability effects of replicates of a single second chromosome, with different numbers of spontaneous polygenic mutants, tested in homozygous and heterozygous conditions. The "coupling" heterozygotes (graph *a*) possess one such chromosome and one "mutant free," or "normal viability," chromosome ($+_i/+_0$), while the "repulsion" heterozygotes (graph *b*) carry two dissimilar mutant chromosomes ($+_i/+_j$). The "coupling" heterozygotes are seen to perform better than the "normal viability" homozygotes ($+_0/+_0$), with all points being above the control value ($+_0/+_0$), set at 1.0, while the "repulsion" heterozygotes are all inferior. The overdominance expressed by the "coupling" heterozygotes reaches an optimum level corresponding to a homozygous viability of between 0.3 and 0.5 (chromosomes yielding these values are estimated to carry about 8 to 11 mutant polygenes). The viability values of the homozygotes and heterozygotes (X and Y axes) were adjusted for each of the two graphs (*a* and *b*) with reference to the performance of the standard homozygote, $+_0/+_0$. With these adjustments, the regression coefficient, 0.39, in graph *b* can be taken as a measure of the average degree of dominance, $\bar{h}$, of newly arisen mutant polygenes. From T. Mukai, *Proc. 12th Intern. Congr. Genet.*, Vol. II (1968), 159–60.

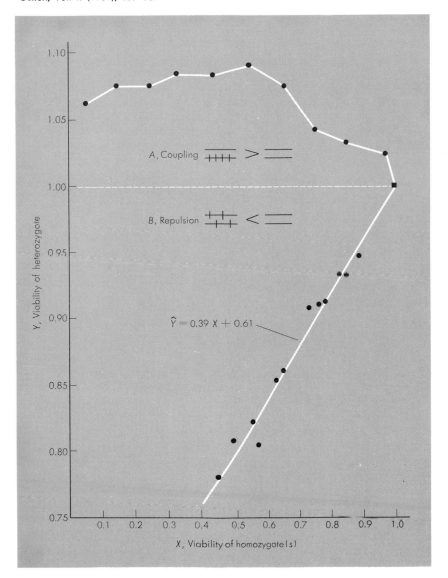

(1960). Wallace found that irradiated chromosomes usually had lower homozygous viability than normal but higher heterozygous viability and thus were overdominant. Wallace's finding was surprising, but Greenberg and Crow discovered that this did not hold true for chromosomes in natural populations. In the latter case the chromosomes with lower than normal homozygous viability also had lower than normal heterozygous viability. Thus they showed a slight degree of dominance, but were not overdominant. As it turns out, the methods of the two experiments were such that Wallace was testing heterozygous viability in "coupling phase," whereas most of those tested by Greenberg and Crow were in "repulsion phase." Thus the discrepancy arose.

It has been pointed out by others that even if the genes at different loci were functionally independent, homozygosity at more than one locus would increase the likelihood that a given individual would die. Thus many homozygous loci (or much of the load) can be eliminated by a single death.

Still another possibility is that many of the polymorphisms may be maintained in natural populations because each genotype is actually superior in its particular subniche in the environment.

For the past several decades, the central problems concerning the genetic architecture of natural populations, and indeed one of the central problems of genetic-evolutionary theory itself, have been those related to: (1) how much variability there actually is in a diploid, random-mating population; (2) what proportion of the loci of an average individual are heterozygous; (3) the genetic burdens associated with the variability; (4) how the variability is maintained in the face of these loads. Although our knowledge is far from complete with regard to all of these problems, it is becoming clear that the answers to the first three questions are far higher values than we had expected. Further progress in this area may result in some surprising changes in our ideas about the genetic architecture of populations.

The Ludwig effect, disruptive selection, and interdemic diversification

Another major factor contributing to the so-called balanced load results from those balanced polymorphisms maintained by disruptive selection. This involves simultaneous selection for two or more genotypes. By selecting for both homozygotes it is possible to maintain a high frequency of both alleles in a population even if the heterozygote is inferior. This gives rise to diversification shown especially well in old established populations in the central regions of the species range, and is often called the "Ludwig effect." Ludwig (1950) argued that a genotype capable of utilizing an unoccupied subniche could survive even if it were less fit than the average in the parts of the niche utilized by other genotypes. The fact that many species tend to be much more polymorphic, both morphologically and chromosomally, in the center of the range, and much less so at the margins (Carson, 1958), lends support to this concept. Presumably, certain basic conditions are more favorable for the species in the central portions of its range. This being the case, it can persist in those parts of the environment which are favorable, until it is able, gradually, to occupy new niches by developing a number of polymorphisms (by disruptive selection) that enable it to utilize its environment ever more efficiently. In marginal areas, the basic conditions are much less favorable and the populations that have managed to invade them are narrowly adapted to one or a few subniches in an otherwise hostile environment. Disruptive selection extending to different geographical areas will lead to interdemic diversification and race formation (Chapter 8). Given time, marginal demes too will tend to become more highly diversified intrademically if they do not die out.

With regard to polymorphisms in general, one should be careful not to be too rigid in defining them, nor too arbitrary in categorizing them in terms of the mechanisms by which they are maintained. It is quite likely that some polymorphisms are maintained by a combination of the mechanisms previously listed. For instance, overdominance and the Ludwig effect are not mutually exclusive mechanisms, and they probably complement each other in many situations. Some evidence for this comes from the observations of Heuts (1948), on the Standard and Chiricahua inversions in *D. pseudoobscura*. At temperatures of 28° to 30°C., the average longevity of the *St/St* homozygotes was greater than those of either the *Ch/Ch* homozygote or the *Ch/St* heterozygote. However, at 0° to 4°C. the longevity of the heterozygotes was greater than that of either homozygote. So we can conclude that, under some environmental conditions (subniches) even in the wild, the homozygotes are superior, even though the same inversion is also more or less preserved by overdominance in other parts of the niche, or under other conditions. Thus the polymorphism can be maintained through disruptive selection in conjunction with overdominance.

Another interesting way in which balanced polymorphisms can be maintained is by frequency-dependent selection. This simply means that a gene is selected for when it is rare, and is selected against when it is common. In other words, its fitness depends on its frequency. We see examples of this in some aspects of mimicry (refer again to Figs. 5.5 and 5.6). When the genes for the mimic form in a polymorphic species, as when one form mimics, say, a highly distasteful model and when the frequency of mimics is low, then there is selection favoring the allele producing it. This occurs because the sheer scarcity of the mimic trains the predators especially well to "recognize" them as the model. Subsequently, as the frequency of the allele and the number of mimics increase, the predators do not learn to avoid them as well, so that they may come to be selected against.

A more thoroughly documented case of frequency-dependent selection, which in addition is probably more general, has been discovered in *Drosophila melanogaster* by Kojima and co-workers (1967). They have studied the esterase-6 locus which controls the production of a particular esterase enzyme. There are two alleles at this locus, F and S, which produce fast and slow electrophoretic variants of the enzyme molecule respectively (Fig. 7.5). Starting with a large laboratory population in which the frequencies of F and S had been stabilized at 0.3 and 0.7 for 15 generations, they did the following experiments. First they extracted a number of lines from the base population, by making single pair matings, which were homozygous for the F or the S allele. By making crosses among these lines they found that the FS females laid more eggs on each of two laboratory culture media than FF or SS females, and that FF females laid more than SS. Average counts for the three genotypes on cornmeal medium were: FF, 49.83 ± 1.65; FS, 52.97 ± 1.62; and SS, 46.71 ± 1.65, for a 4-day period. Thus there must be other components of fitness which compensate (remember, the frequency of S in the base population is 0.7) for the lower fecundity of SS. To test this egg-to-adult viability, tests were carried out, with remarkable results. As before, crosses were made between the homozygous FF and SS

Fig. 7.13. Allelic frequency changes on esterase-6 locus (*D. melanogaster* on banana medium). Figures 7.13 and 7.14 show frequency-dependent selection at the esterase-6 locus in cage populations of *D. melanogaster*. The point to be stressed is that in cages in which the initial value of F is high, the frequency falls and when it is low, it increases. The adaptive value of F depends on its frequency. From K. Kojima and K. Yarbrough, *Genetics, 57* (1967), 677–86.

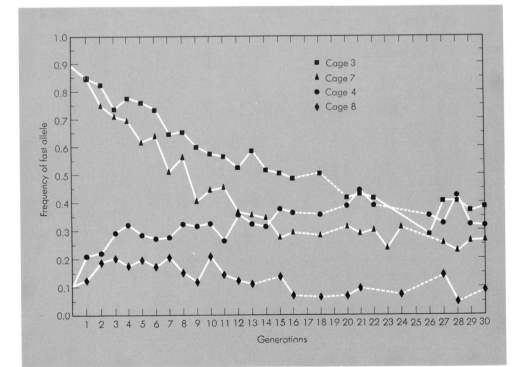

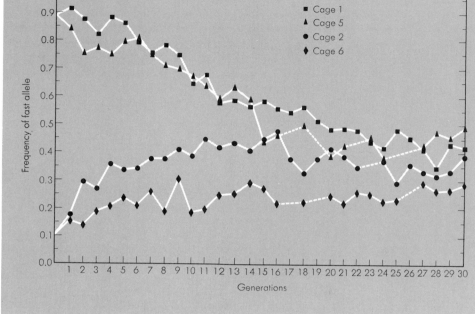

Fig. 7.14. Allelic frequency changes at esterase-6 locus (*D. melanogaster* on cornmeal medium). From K. Kojima and K. Yarbrough, *ibid*. (see Fig. 7.13).

lines and the fertilized females were put into population cages where they could lay eggs. Females were put into different cages in groups in such a way that they represented Hardy-Weinberg populations with different gene frequencies. The expected proportion of each genotype was calculated from the Hardy-Weinberg equation, while taking into account differences in fecundity of the *FF*, *FS*, and *SS* females. Their four cages were as shown in Table 7.3.

Table 7.3.   Expected proportion of *FF*, *FS*, and *SS* genotypes

Frequency of F	FF	FS	SS	Totals
0.7	0.511	0.393	0.096	1,000
0.5	0.260	0.497	0.243	1,000
0.3*	0.105	0.403	0.492	1,000
0.15	0.047	0.250	0.703	1,000

* Same as base population.

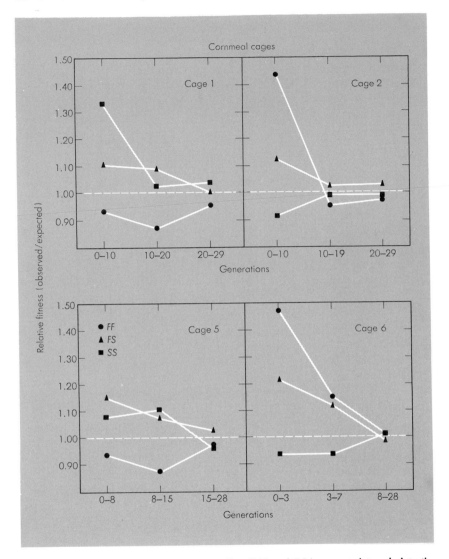

Figs. 7.15 (*left*) and 7.16 (*right*). The data in Figs. 7.13 and 7.14 were used to calculate the relative fitness values shown here. This shows that when the population is far from its final equilibrium (generations 1–10) the adaptive values for *F* and *S* are high or low, depending on whether their values are above or below their equilibrium values. At equilibrium, however,

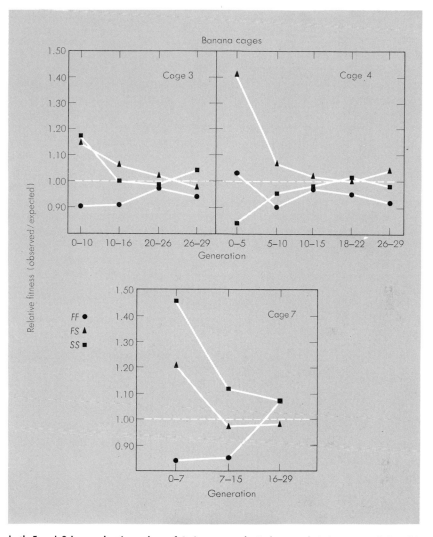

both *F* and *S* have adaptive values of 1, i.e., are selectively neutral. It is suggested that this is one reason many polymorphic loci can be maintained in a population without incurring a disastrous genetic load. From K. Kojima and K. Yarbrough, *Genetics, 57* (1967), 677–86.

However, when the offspring were examined they found that these expectations were not always met. The viability estimates (observed proportion/expected proportion) were as shown in Table 7.4.

Table 7.4.   Viability estimate and their standard error

Frequency of F	FF	FS	SS
0.7	0.759 ± 0.048	1.140 ± 0.061	1.713 ± 0.151
0.5	0.842 ± 0.078	1.034 ± 0.049	1.099 ± 0.084
0.3	1.148 ± 0.151	1.016 ± 0.062	0.954 ± 0.052
0.15	1.677 ± 0.195	1.726 ± 0.075	0.697 ± 0.028

Note that when the observed proportion equals the expected proportion the viability estimate is 1. Strikingly, when the frequency of F is above the equilibrium of the base population the viability of FF is much lower than expected, and the viability of SS is much higher. This is especially true of the cage in which $F = 0.7$. On the other hand, when the frequency of F is below that in the base population, FF is more viable than expected and SS is less so. This indicates clearly that the relative viabilities among genotypes depend upon the gene frequency. In this case, F is selected for when it is below the equilibrium frequency and is selected against when it is above equilibrium frequency. The reverse is true for S. At the equilibrium value the viability of all three genotypes is very close to that expected. This same phenomenon can be observed in Figs. 7.13 and 7.14. When population cages are started with either high or low initial frequencies of F, it can be seen that by the end of 30 generations the frequency of F has either increased or decreased to about 0.3. Again, when F is high it is selected against, and vice versa. In Figs. 7.15 and 7.16 we can see that the viability estimates (equated to fitness estimates) are higher or lower than expected in early generations, but that they aproach unity (expected) in later generations. Figures 7.15 and 7.16 were based on the data from the cage populations in Figs. 7.13 and 7.14. From these data it appears that at equilibrium all three genotypes are selectively neutral with respect to viability. By definition there is no genetic load at this locus, because all genotypes have equal fitnesses. They conclude that this is a possible general mechanism responsible for the large amount of genetic polymorphism in nature.

Another type of frequency-dependent selection has been reported in *Drosophila* by Ehrman (1966), that depends on mating behavior. In several experiments in which mating propensities of individuals with different genotypes were observed, it happened that individuals with rare genotypes mated much more effectively than individuals having the same genotype who belonged to populations in which the particular genotype was common.

At any rate, we can see that overdominance is not the only means of maintaining polymorphism in a population, and consequently that balanced loads are not due entirely to overdominant loci.

Directional selection, transient loads, and the cost of evolution

One can also single out a particular kind of load in cases of dynamic gene frequency changes such as occur when one allele is in the process of being replaced by another in the population. This process, called "gene replacement," is one of the basic features of evolution.

Such a replacement is usually initiated by a change in the environment which results in a shift in the direction of selection so that a previously unfavorable allele is now favored, and vice versa. For the duration of the shift, this type of selection is referred to as "linear" or "directional" selection to distinguish it from balancing selection, or normalizing selection. This will cause the frequency of one allele to decrease as the other increases. For a time during this period of transition there will be a high frequency of both alleles. Consequently, there will be a transient period of polymorphism (in cases of genes with major phenotypic effects) accompanying this shift. Of course, when selection pressures shift so that the common allele is selected against, a large proportion of the population necessarily will be ill adapted, and this places a load on it which is also transient, and which is referred to as the "substitutional," or "transient" load.

Haldane (1957) has made some very interesting calculations of the cost to a population of making an allele substitution. According to his calculations, the number of deaths, or their equivalent in reduced fertility, caused by a shift in selection is virtually independent of the intensity of selection, and depends only on the initial frequency of the previously unfavorable allele. However, as one would expect, the number of generations over which the deaths must be spread, in order to make the shift, is directly determined by selection intensity. The relation between selection intensity and the number of generations over which the deaths occur, for an average locus ($q = 0.005$), is approximately $I = 30/n$, where I = selection intensity, n = the number of generations, and 30 = the factor by which the total number of deaths over generations exceeds the number of individuals existing in any given generation. If I is 0.1, an amount many people consider as a reasonable average estimate during gene replacement, then $n = 300$. It is estimated, then, that an average gene replacement requires 300 generations and 30 times as many deaths as there are individuals in the population each generation![1] How many substitutions can be made simultaneously without reducing the fitness of the population to an impossibly low level? Haldane, using much the same reasoning as was applied to determining the limits of the number of overdominant loci, has argued that the cost of evolution in terms of gene replacement must be very high. Kimura (1960) has reached a similar conclusion. For instance, in the example of industrial melanism in the moth *Biston betularia,* when the peppered moths are experimentally introduced into the industrial areas, predation by birds reduces their numbers by half each day. So if we take 0.5 as the intensity of selection at this locus, 10 such loci would reduce the survival rate to 1/1,024, or $(\frac{1}{2})^{10}$. In some species, this might be tolerable if the reproductive rate was sufficiently high, but in many it would lead to extinction. Then, too, how many genes must be substituted to enable a population to remain in a changing environment, or to become better adapted to a new niche? It is probably many more than 10. However, Haldane argues that this is, nonetheless, a very slow process and increasing the number of loci that are simultaneously involved by decreasing the intensity of selection (which is to say, by decreasing less drastically the fitness) of each one, does not help the situation. This simply increases n, the number of generations required per substitution. The average number of substitutions still is 1 per 300 generations.

Whatever the cost of gene replacement, as Brues (1964) points out, the cost of *not* evolving is higher. When the environment changes, selection for those rare individuals with higher fitness will improve the fitness of the population, although it is costly to have so many ill-adapted individuals eliminated.

Rates based on Haldane's calculations fit fairly well with the observed times required for changes giving rise to new species in the fossil record. If one assumes, as Haldane did, that 1,000 gene substitutions are sufficient to give rise to a new species, it would take at least 300,000 generations. For many mammalian species, this works out to be about the right length of time to conform to the time periods thought to have been involved, judging from paleontological evidence.

On the other hand, there appear to be many instances where the

---

[1] This number **30** is calculated from the following equation:

$$D = \frac{1}{1-\lambda}\left[-\ln p_0 + \frac{\lambda \ln(1-\lambda)}{\lambda}\right]$$

$D =$ deaths, when the genotypes, fitnesses, and frequencies are as follows:

Genotype	AA	Aa	aa
Frequency	p^2	$2pq$	q^2
Fitness	1	$1-k$	$1-K$

$k = \lambda K$. Thus when $\lambda = 0$, A is completely dominant, and D becomes simply $-\ln p_0$. For a more complete explanation of this equation and its derivation, see Haldane (1957).

rates of evolution are higher than could be accomplished by 1 substitution per 300 generations. In the case of man, we have evolved from some *Homo erectus* line within the last million years. If we take the average generation time as 20 years (perhaps a little high) we have made an enormous advance in cranial capacity, and associated morphological and physiological features, in a mere 50,000 generations. At the rate of 1 substitution per 300 years, there is enough time to allow for only 166 gene changes. This seems far too low to account for the magnitude and complexity of the changes that have occurred. Many other cases of rapid evolution could be cited.

Other, perhaps more likely, possibilities have been suggested to explain the apparent discrepancy between the low rate of gene substitution and the occasional high rate of evolution. Mayr (1963) has suggested the theoretical possibility that the reduction in fitness a different loci is synergistic, i.e., the reduction in fitness due to having an undesirable genotype simultaneously at two loci is more than twice that attributable to either locus separately. This means that those individuals whose death would eliminate undesirable genes at several loci are much more likely to die, and therefore the potential genetic deaths are concentrated in fewer individuals. Another possible factor in reduction of fitness could be a markedly decreased intraspecific competition that results from the reduction in numbers.

Whatever the true explanation, it should be interesting to follow developments in the next few years that will aid our understanding in these areas.

Homeostasis and canalization

In view of the large amount of genetic variation that is present in most Mendelian populations, and the vast number of possible genotypes, it is to some degree surprising that the normal, or "wild-type," phenotype is expressed so uniformly. To be sure, there is much more variability than meets the eye of the casual observer, but still, the fact remains that a vast number of genotypes give rise to phenotypes which are indistinguishable, even under diverse environmental conditions. The explanation of this phenomenon is not fully understood, but part of the answer, at least, lies in the concepts of developmental homeostasis and canalization. These are to some extent overlapping terms. Developmental homeostasis refers simply to the general phenomenon whereby a given genotype gives rise to the same phenotype over a wide range of environmental conditions. Canalization is somewhat more specific than this and refers to genetic interactions that progressively narrow the possible sources of phenotypic expression as the individual develops.

Developmental homeostasis should not be confused with genetic homeostasis, which is a term used to describe the resistance of a population to changes in its genetic composition by selection. This resistance to change, or at least to rapid change, is in large degree due to the fact that an organism is a highly integrated, delicately balanced entity, and to change part of it is to upset the balance of the integrated geno- type, giving rise to individuals with low fitness. Although genetic homeostasis undoubtedly involves the integration of developmental processes, it is not the same thing as developmental homeostasis, which is an individual phenomenon. Genetic homeostasis is thoroughly discussed by Lerner (1954).

The concept of canalization has been developed largely by C. H. Waddington of the University of Edinburgh. He has asserted that stabilizing selection consists of two more or less distinct aspects. First, there is "normalizing" selection, which results in the actual elimination of abnormal phenotypes, and "canalizing" selection, which eliminates individuals whose development is variable, or erratic, and poorly buffered against environmental fluctuations. Canalization is envisioned as involving major controlling genes, sometimes called "switch genes," which stabilize the phenotype both against genetic variation at other loci, and against environmental variation. In the case of sexual differentiation in higher animals during a certain period in development, for instance, certain major genes begin to function that will determine the sex of the individual. This somehow sets the limits within which the *many* other genes that are concerned with the development of maleness or femaleness are constrained to act. In other words, there will be many genes determining the phenotypic expres- sion of the secondary sex characters, but in spite of the potential variability inherent in these many genes, the final result is highly uniform. It would require the cumulative effect of a great many loci to transgress the limits set by the major "switch," or "canalizing," genes.

Many authors have concluded that if a particular phenotype is ad- vantageous, there will be strong selection for genes canalizing its devel- opment, so that the developmental pathways in virtually all organisms are highly canalized. Support for this point of view comes from two major sources. (1) In many different kinds of organisms, there are large numbers of sibling species which on morphological grounds are virtually indistinguishable. It appears that in such cases highly canalized developmental patterns have permitted a fairly extensive alteration and reorganization of the genotype without changing the phenotype. (2) Most wild-type organisms are fairly resistant to arti- ficial selection for a wide variety of morphological traits. For instance, in *Drosophila* it is hard to select, from a wild-type strain, a strain with an altered number of scutellar bristles or of eye facets. It is thought that this is because the bristle number and the facet number result from canalized developmental pathways. However, selection for strains with altered bristle number or number of eye facets in mutant strains, such as "scute" or "Bar," is quite successful. Scute is a mutant character that results in the complete absence of four particular thoracic bristles, and this character affects other bristles and hairs to some extent. The success in selecting for more bristles, in addition to the fact that scute flies themselves are variable in the degree to which the bristles are affected, indicate that the wild-type allele plays a canalizing role which is disrupted in scute. The same general type of result occurs with Bar eye. This mutant produces different numbers of eye facets under different temperature conditions. At 25°C., the average is 55 facets, and at 18°C., 156 facets. The normal eye contains about 750 facets. Presumably, canalization has been disrupted in Bar and facet development has become temperature sensitive. But the same disruption renders selection effective so that the difference in number of eye facets between flies developing at 18°C. and those at 25°C. can be reduced from 300 percent to 10 percent. Another situation, quite like the scute example, involves the coat texture mutant, "Tabby" in mice, which also affects the whiskers. In wild-type strains, the whisker number is quite constant and resistant to selection. In Tabby strains, it is possible to select for higher or lower numbers of whiskers. Canalization obviously is not a phenomenon retricted to *Drosophila*. These experiments have been carried out by Waddington (1950), Rendel (1959), and Dun and Fraser (1958).

In considering the contributions of individual loci to developmental homeostasis, we should point out again that heterozygosity itself, which on the one hand constitutes the variability in many instances, must result in greater phenotypic stability because alternate forms of a polypeptide can be produced, with each serving the needs of the organism more satisfactorily under different conditions.

Recombination and genetic systems

One of the most important aspects of the genetic architecture of a population is the genetic system of the species and the role of recombination in the production of new genotypes. By "genetic system" we mean the manner in which the genes are organized within the individual and the manner in which they are transmitted to offspring. The genetic system involves such things as number of genes, number of chromosomes, type of chromosome structure, ploidy, average mutation rate, crossover frequencies, details of meiosis, type of breeding system, type of sex determination, methods of regulating gene activity, and so forth. Genetic systems, as Grant (1958) has pointed out, vary considerably from species to species and, to a lesser degree, within each species. Since the components of the genetic system are themselves under control of the genotype, they are subject to natural selection. As a result, the genetic system is no more invariant than the morphology or physiology of the organism and, through the action of natural selection, will become adapted to the environmental conditions with which the organism must cope. In other words, genetic systems are subject to the same kinds of evolutionary divergence, convergence, and parallelism as other characters. Darlington and Mather pointed this out in the late 1930's, but it is a fact that had received little attention until recent years. Many features of genetic systems are universal and undoubtedly are so because of their common descent from an original type. For instance, all available evidence indicates that the RNA triplets (codons—see Philip E. Hartman and Sigmund R. Suskind, *Gene Action,* in this series) which specify particular amino acids are the same in all organisms. The components of the genetic control of protein synthesis are interchangeable, in cell-free extracts, among such diverse organisms as the bacterium *E. coli,* pea seedlings, and rabbits. Although they are interchangeable and utilize the same RNA code words, they are not identical, as Spiegelman (1966) has shown by his elegant nucleic acid complexing experiments. Other features, such as the presence of chromosomes and occurrence of meiosis, are widespread, but not universal, and conceivably could have arisen more than once in the course of evolution.

Our tendency to consider genetic systems as relatively fixed, unchanging entities has produced several surprises, as our knowledge has extended to a greater number of species. Sex determination in *Drosophila* involves a balance of genes determining maleness and femaleness distributed strategically on the X chromosome and the autosomes. The Y chromosome is not involved, except in a very minor way, in that it has a few genes for sperm motility. Thus, a zygote lacking an X chromosome (XY or XO, with the O indicating the complete absence of a second sex chromosome) develops into a male, whether it has a Y chromosome or not. Individuals with two X chromosomes, XX or XXY, develop into females whether there is a Y chromosome present or not. Evidently, the absence of an X (and of course its female-determining genes), rather the presence of a Y, determines maleness. The sex chromosome constitution of normal female *Drosophila* is thus XX, and of normal males XY. This is also true of mammals, including humans. In the mammalian genetic system, however, the Y chromosome determines maleness. Individuals with the abnormal chromosome constitution XXY are males rather than females, and XO individuals are females (see Victor A. McKusick, *Human Genetics,* in this series). Another surprise involved tolerance of aneuploidy (a condition in which there are extra or missing chromosomes) of the X chromosome. In *Drosophila* females, two X chromosomes are normal, but the presence of three produces highly abnormal individuals which rarely are viable. These females are never fertile, and usually fail even to survive the pupa stage. These are called inappropriately "superfemales" because of the extra X chromosome which, in addition to contributing more female-determining genes, upsets the genetic balance. The first report of superfemales in humans, consequently, was accorded much skepticism. But, as it turns out, there is again a difference in the genetic systems of the two species, for in humans and in other mammals, all X chromosomes in excess of one become largely inactivated at an early stage in embryonic development.

Having established that genetic systems are subject to natural selection and, predictably, are not uniform, let us turn our attention to a feature of genetic systems that is universal, i.e., recombination.

Recombination occurs from the simplest viruses to the most complex multicellular organisms and it is the essence of sexual reproduction. Whether or not the mechanisms that result in recombination are the same in all organisms is not at issue here, although this too is an interesting point. The question, as it concerns evolution, is not whether recombination occurs, but how extensive it is. Clearly, free recombination in every generation among all genes would be utterly chaotic, so far as natural selection is concerned. This problem, along with the obvious difficulty in achieving the regular distribution of a large number of elements during cell division, are two compelling reasons for the need to organize the genes into larger units, the chromosomes. If recombination is to be retained, on the other hand, there must be some means of reshuffling the genes that are in the same chromosome. In all organisms, this is accomplished by a process called "crossing over," which seems to involve a precise breakage and reunion, always of homologous chromosomes. (See Carl P. Swanson, Timothy Mertz, and William J. Young, *Cytogenetics*, this series.) In viruses, this seems to be a rare occurrence, because the requisite simultaneous presence of two different kinds of virus in a single cell is rare. Among bacteria, donor cells and recipient cells occasionally will conjugate and engage in partial or complete chromosome transfer, and recombination thereby can occur in the recipient cell. Bacterial conjugation, however, appears to be rather rare in nature. In higher organisms, crossing over between homologous chromosomes is the rule. The number of crossovers per chromosome pair depends on a number of factors, but basically is determined by the size of the chromosome. Very small chromosomes may never cross over, whereas larger chromosomes will average one or more per meiotic division. Of course, since chromosome pairs segregate independently in most organisms, genes on different pairs of chromosomes are still subject to free recombination. Judging from its universal occurrence, recombination even of linked genes (genes on the same chromosome pair) must have a very strong selective advantage. This is easy to understand, in view of the existence of integrated gene complexes and the interaction between genes that we have considered. If a new mutation were the only way in which a new combination of genes could arise, the process of producing and testing new genotypes would be impossibly slow, unless mutation rates were intolerably high, and integrated complexes never would be constructed. However, if mutations that have arisen in different individuals can be included in the same chromosome by recombination in heterozygotes, the efficiency of producing new genotypic combinations is tremendously enhanced. There are a number of ways, however, in which crossing over, and thus recombination, is restricted. Presumably, the reason for this is that although recombination is essential for the construction of an integrated genotype, too much recombination is detrimental to the retention of integrated complexes—too much recombination breaks them up. Examples of restriction on recombination include inversion heterozygotes in *Drosophila* and other insects, the restriction of crossing over to one sex, parthenogenesis altering the amount of heterochromatin in a chromosome, or physiologically reducing the amount of crossing over that occurs, and inbreeding. Furthermore, the amount of recombination that is advantageous seems to vary. Again citing some examples from *Drosophila,* in *D. melanogaster* and *D. hydei* the X chromosomes are about the same physical size. The genetic map of the X in *D. hydei,* however, is twice as long as that of *D. melanogaster.* Since the construction of genetic maps is dependent on crossing over, the frequency must be twice as high in *D. hydei.* Also, as Carson (1958) has shown, inversion polymorphism is much more pronounced in central regions of a species range than in the marginal areas. Thus, the breakdown of integrated complexes, highly adapted over long periods of time, appears to be disadvantageous in the central regions. In newly inhabited marginal areas, the highly adapted, integrated complexes have not yet developed and high frequencies of recombination are more advantageous for the possible development of such complexes, and for producing large numbers of new genotypes, to be tested in the relatively new environment.

In this chapter we have seen that there is a large amount of genetic variation in most populations, hidden behind a facade of normality. We have also seen some of the ways in which it is preserved, some of the costs of preserving it, and some of the ways of utilizing it. And finally, we saw that the assumption that "different loci act independently on fitness" is being made in areas in which it is critical, and that whatever its usefulness in other areas, it will probably have to be modified for determining the limits of variation that are possible in a population.

References

Brues, A. M., "The Cost of Evolution Versus the Cost of Not Evolving," *Evolution, 18* (1964), 379–83.

Carson, H. L., "Response to Selection Under Different Conditions of Recombination in *Drosophila,*" *Cold Spring Harbor Symposia on Quantitative Biology, 23* (1958), 291–306.

Crow, J. F., "Some Possibilities for Measuring Selection Intensities in Man," *Human Biology, 30* (1958), 1–13.

Dobzhansky, Th., *Genetics and the Origin of Species.* New York: Columbia University Press, 1951.

Dobzhansky, Th., "Evolution as a Creative Process," *Proceedings of the 9th International Congress of Genetics* (1954), 435–49; reprinted in *Papers on Animal Population Genetics,* Eliot B. Spiess, ed. (Boston: Little, Brown & Co., 1954).

Dobzhansky, Th., "How Do the Genetic Loads Affect the Fitness of Their Carriers in *Drosophila* Populations?" *American Naturalist, 98* (1964), 154–66.

Dobzhansky, Th. and B. Spassky, "Genetics of Natural Populations XXXIV.

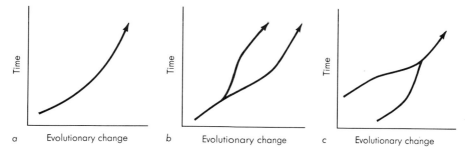

Fig. 8.2. (a) **Phyletic evolution.** (b) **Primary speciation.** (c) **Secondary speciation.**

tion), which results in a complete redefinition of the adaptive field itself. The peaks and valleys will change as different genotypes become more or less fit.

If changes occur throughout the species range the entire species will be affected more or less uniformly. This is what we have called phyletic evolution. It is often classified as a mode of speciation because it can produce new species, but as we have pointed out, it does not increase the number of species in existence.

When different populations within a species are subjected to different environmental conditions, different kinds of changes will occur and, if their magnitude is sufficient, one species can split into two or more contemporary species. We have referred to this process as speciation, and it is another mode whereby new species can arise.

Yet another possibility exists for the formation of new species. This involves the fusion of two existing species to form a single new one. This has sometimes been called secondary speciation. These three modes of speciation are represented schematically in Fig. 8.2. The extent to which secondary speciation occurs is a subject of some debate, but it is undoubtedly more important in plants than in animals because of the much greater ease of obtaining crosses between different species.

In each of these modes we can recognize our oft-repeated theme of natural selection operating on existing variability to bring about evolutionary change. But let us further consider speciation in its more restricted sense, i.e., in the proliferation of species.

Race formation

The first step of speciation requires that different populations in a species be subjected to different environmental conditions. In the process of becoming adapted to these, the populations themselves will become genetically different. This divergent phyletic change constitutes race formation among populations and is the essence of speciation. As the result of accumulating genetic differences, distinguishable

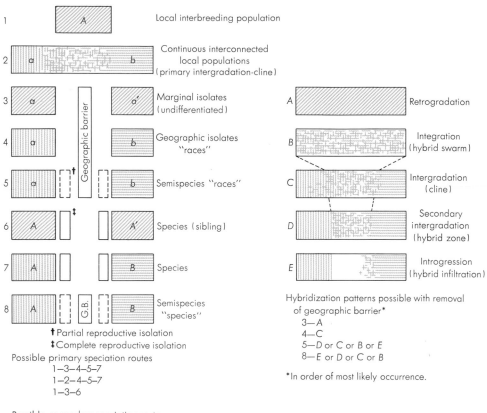

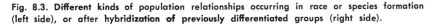

Fig. 8.3. Different kinds of population relationships occurring in race or species formation (left side), or after hybridization of previously differentiated groups (right side).

morphological and physiological differences will follow. When populations have accumulated enough changes of this type to make them distinctive they are usually recognized as races, and may be formally named as subspecies. This is the first recognizable step in speciation. *The populations have become genetically differentiated in the process of adapting to different environments.* By definition, races still are potentially interfertile, with no reproductive isolation having arisen. Their fate undoubtedly is variable and depends on a number of factors. If the environmental differences persist or become more pronounced, the races may become more divergent until they have become so different genetically that they can no longer interbreed freely, at which point we consider them to be separate species. On the other hand, the races may persist more or less indefinitely or, by the resumption of gene flow, meld back into a single, interbreeding population. Various relationships between populations in race formation and species formation are depicted schematically in Fig. 8.3.

It is important to recognize that the gradual and continuous nature of race formation and species formation makes it impossible to draw distinct demarcation lines and to make precise definitions (see Chapter 2). At what point does a homogeneous species become subdivided into races? At what point do races become distinctly new species? The history of biology is filled with controversies which still continue today over the proper classification of various groups. Attempts to categorize different degrees of divergence have led to the use of such terms as "race," "geographical race," "ecological race," "variety," "local variety," "subspecies," and others. Some authorities have used "variety," "race," and "subspecies" to denote progressive levels of divergence. However, in the absence of any real distinction between these categories, they have come to be used more or less as synonyms, according to the preference of each author (the term "subspecies" is the formal taxonomical category). It now is clear that most of the controversies over classification stem from an attempt to impose discontinuous categories on a continuous process. There are, of course, stages in the process, the identification of which can easily be agreed upon. There are times when a particular species is clearly composed of distinct races, or when two greatly different groups, such as tigers (*Felis tigris*) and mountain lions (*Felis cougar*), constitute different species. But these clear-cut stages develop gradually through a progression of less clear-cut steps.

Isolating mechanisms and speciation

Isolating mechanisms are of paramount importance in speciation and so have been extensively studied. For a more thorough treatment of the subject consult Dobzhansky (1951), Patterson and Stone (1952), Grant (1963), and Mayr (1963).

Speciation, as we have seen, requires the development of reproductive isolation between populations. But before reproductive isolation can become established, it is necessary to first have some sort of physical separation to initiate restriction to gene flow. Perhaps the most common type of physical separation is geographical. That is, the individuals simply inhabit different geographical areas.

Sometimes geographical isolation involves nothing more than separation by distance. Frogs in North Carolina cannot mate with frogs in Ohio, nor Maine robins with Michigan robins. When the frogs in North Carolina reproduce, they will be subjected to selection on the basis of local environmental conditions which, in many ways, will differ from those in Ohio. This often will give rise to a species in which the phenotypes in one area gradually and imperceptibly grade into those typical of other areas, giving rise to a cline. This is because geological, climatic, and other features of the environment within the species range also show the same gradual changes. Furthermore, when some features of the environment are discontinuous, the gene flow between adjacent, contiguous populations will often tend to obscure them. This type of geographical isolation is well illustrated in the leopard frog, *Rana pipiens*, a species distributed over the entire eastern half of the United States. Starting any place within their range and striking out in any direction, sampling at closely spaced intervals, at no place does one discover a line of discontinuity between one form and another. When specimens from more widely spaced areas are examined, morphological distinctions become apparent. Finally, in testing the Texas frogs and Vermont frogs for interfertility, it was discovered that they were, to a large extent, isolated, even though closely adjacent populations are interfertile throughout the cline (Moore, 1949). Actually it now appears that in their natural habitats certain ones of these races are isolated on the basis of differences in mating calls. This would necessitate their reclassification as species. But the fact remains that isolation by distance can give rise to races even though there are no sharp lines of demarcation between them [Fig. 8.3(2)]. Such races would have to be classified as allopatric.

In many other cases there are more or less sharp breaks in the environmental conditions within a species range, and correspondingly sharp lines of demarcation between the various geographical races [Fig. 8.3(4)]. Such is the case with the song sparrows in the San Francisco area, where there is an upland race and a lowland race restricted to the salt marshes. This is also true of several races of pocket gophers in the state of Washington, each being restricted to a different soil type. In these cases, too, the races are classified as allopatric. However, it is essential to recognize that such races can be nearly or entirely disjunct, as they will be when the ecological factors on which a species depends are distributed in an insular fashion, or are contiguous, as when the different sets of environmental factors are contiguous.

Many other factors can result in geographic isolation. The rise and fall of land masses such as the Bering Strait or the Isthmus of Central America, have been very effective in isolating land animals or sea animals. Simpson (1950) has dealt with the very interesting developments in the Latin American fauna resulting from changes in the Central American land bridge which was submerged during the tertiary period of the Cenozoic era. Such things as glaciers or changes in climate also can very effectively result in geographic isolation. The evidence is strong that alligators and magnolia trees once inhabited a range extending from southeastern China to southeastern United States, along with other species as well. Their range was simply cut in half by Pleistocene glaciation. Furthermore, any of these species which were clinally distributed, as in our frog example, could have had the races at either end of the range elevated to unequivocal species status simply by the elimination of the intermediate populations. The formation of a river, a mountain range, a desert, or the occurrence of forest fire can all result in geographic isolation. Yet, as important as these major events may be at certain times, especially for long-term isolation, they don't occur often enough to account for the tremendous number of species which are subdivided into races. Much more important is the discontinuous distribution of ecological factors.

As discussed in Chapter 5, Mayr (1963) makes a strong case for geographic species formation, through the further genetic divergence of allopatric races, being the predominant way that speciation can occur; and that most if not all reported cases of sympatric speciation are really cases of secondary sympatry following allopatric speciation.

However, many people still contend that sympatric races do occur and that they can culminate in sympatric species formation (see Grant, 1963). It is alleged that this is most often initiated through the preference of individuals with different genotypes for different subniches in the same geographical area. To the extent that each genotype restricts its activities to its preferred subniche, subpopulations will arise that are physically separated from the others. In this case the physical separation has a genetic basis. This type of isolation is usually called ecological isolation, but it has also been called environmental or habitat isolation. Another point made by opponents of sympatric speciation is that ecological isolation by itself (i.e., without geographical isolation) probably will not last long enough to allow for speciation, though it might persist long enough to produce sympatric races before breaking down.

Some very interesting examples of ecological isolation of sympatric species certainly exist, but whether it follows that they arose from the further divergence of sympatric races is still open to question. A striking example of the ecological isolation of a species of *Drosophila*, *D. pachea*, has been discovered by Heed and Kircher (1965). This particular species is found only in the Sonoran Desert in the southwestern United States and northern Mexico. It breeds exclusively in the stems of the senita cactus (*Lophocereus schottii*) and is quite incapable of reproducing in the laboratory unless a piece of this cactus is added to the medium. Even more peculiar is the fact that the larvae of other species inhabiting the Sonoran Desert are killed or severely inhibited in their growth by addition of the same cactus to their culture vials. So here we have a situation in which a cactus that is an absolute requirement for one species is poison to others. When the substance required by *D. pachea* was isolated from the cactus, it turned out to be a sterol, Δ^7-stigmasten-3p-ol. Several alkaloids also were isolated from the cactus and seem to constitute substances which are lethal to other species, but to which *D. pachea* is tolerant. With what would appear to be very little genetic change, a case of very complete ecological isolation has become established. It is probable that changes at only a few loci were required to produce enzymes that enabled *D. pachea* to convert the new sterol into a form that could be utilized in its own sterol metabolism while losing the capacity to produce enzymes which convert more universal and conventional sterols, and to produce an enzyme that destroys the lethal alkaloid. In a case like this, with a potentially simple genetic basis giving rise to such complete isolation,

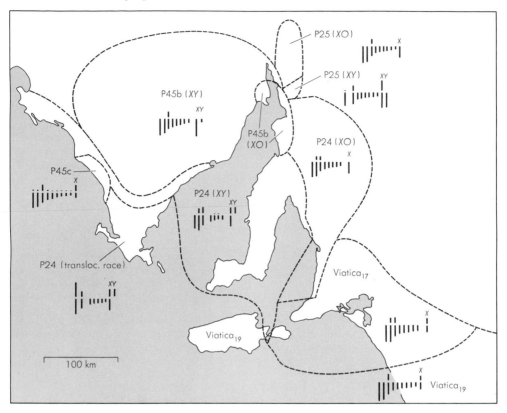

Fig. 8.4. Map of a portion of South Australia, showing the present distribution limits of the members of the "coastal" group of forms of the *viatica* group of grasshoppers. During the Pleistocene glaciation the gulfs between the Eyre, Yorke, and Fleurieu peninsulas were dry land, and Kangaroo Island was also a part of the mainland. The ideograms show the haploid karyotypes of the various forms. In *viatica* the chromosomes from left to right are designated A, B, CD, 1, 2, 3, 4, 5, 6, X. The karyotype of the XO race of P45b is not shown; it is not significantly different from that of the XO race of P25. These species have not yet been named and are provisionally labeled P24, P25, etc. The parapatric distribution of these chromosomally distinct species is not easily accounted for by either allopatric or sympatric speciation: it is called stasipatric. From M. J. D. White, *Science*, 159 (1968), 1067. Copyright 1968 by the Am. Assoc. Advancement Sci.

one is tempted to conclude that this could have occurred sympatrically, perhaps as one possible genotype in a polymorphic population.

Recently White (1968) of the University of Melbourne has proposed another alternative to speciation through the formation of sympatric or allopatric races. He calls it *stasipatric* speciation. In working with the Morabine (family Morabinae) grasshoppers in Australia, White and his co-workers have discovered several species which are parapatrically distributed (Fig. 8.4). There are chromosomal differences between some of these species which are located geographically in such a way as to make it unlikely that these species arose from allopatric races. At the same time, they are not sympatric. The most probable explanation of these observations is that a new species arises within the old species range and takes over whatever area it can by eliminating the old species. This is a different situation than has been previously recognized, and it would have been considered a case of allopatric speciation except for the unusual distribution of chromosome types.

In any case, the kinds of isolation we have discussed, geographic and ecological, are sufficient to interrupt gene flow and to initiate race formation. However, the elevation of races into species usually requires the development of additional isolating mechanisms that result in stronger reproductive isolation.

Reproductive isolation

Let us examine some of the mechanisms that give rise to reproductive isolation (Table 8.1). Various authors have classified them

Table 8.1. Summary of the most important isolating mechanisms

A. *Prezygotic mechanisms.* Prevent fertilization and zygote formation.
1. *Habitat.* The populations live in the same regions, but occupy different habitats.
2. *Seasonal or temporal.* The populations exist in the same regions, but are sexually mature at different times.
3. *Ethological* (only in animals). The populations are isolated by different and incompatible behavior before mating.
4. *Mechanical.* Cross fertilization is prevented or restricted by differences in structure of reproductive structures (genitalia in animals, flowers in plants).
5. *Gamete incompatibility.* Gametes fail to survive in alien reproductive tracts (sometimes called physiological).

B. *Postzygotic mechanisms.* Fertilization takes place and hybrid zygotes are formed, but these are inviable, or give rise to weak or sterile hybrids.
1. *Hybrid inviability or weakness.*
2. *Developmental hybrid sterility.* Hybrids are sterile because gonads develop abnormally, or meiosis breaks down before it is completed.
3. *Segregational hybrid sterility.* Hybrids are sterile because of abnormal segregation to the gametes of whole chromosomes, chromosome segments, or combinations of genes.
4. F_2 *breakdown.* F_1 hybrids are normal, vigorous, and fertile, but F_2 contains many weak or sterile individuals.

SOURCE: G. L. Stebbins, *Processes of Organic Evolution* (Englewood Cliffs, N.J.: Prentice-Hall, Inc., 1966).

somewhat differently but the mechanisms are the same. Prezygotic mechanisms involve either the failure of individuals in the isolated species to meet or, if they meet, a failure to mate. Failure to meet involves primarily seasonal isolation and habitat isolation. Seasonal isolation is common in both plants and animals. The toad, *Bufo americanus,* breeds early in the season, whereas *Bufo fowleri* breeds late in the season (Blair, 1928). Even though they may breed in the same pond, and will cross in the laboratory under conditions of controlled ovulation, they are isolated in nature. In many plants the ripening of ovules and pollen are critical factors in reproduction and are closely controlled by a number of climatic factors. The difference of a week in the time of pollen ripening can result in complete isolation. Numerous other examples could be given.

It is important to point out that a species is not restricted to a single isolating mechanism. For instance, the toads *Bufo americanus* and *B. fowleri* are actually subject to both seasonal and habitat isolation. The latter species not only breeds somewhat later in the season but also shows a distinct preference for ponds, large pools, and quiet streams. The *B. americanus* prefer small shallow puddles and small pools in drying creeks as a breeding site. Nevertheless, their breeding seasons do overlap slightly and they do sometimes breed in the same pool, so that occasional hybrids are formed. On the whole, though, we see here two isolating mechanisms working simultaneously. Differences in mating call have been shown (Blair, 1958) to be yet another isolating mechanism in toads. Furthermore, the mechanisms isolating species A from species B are not necessarily those isolating A from C.

The major mechanism by which potential mates meet but fail to mate is ethological isolation. Ethology is the study of comparative animal behavior under natural conditions. Ethological isolation, therefore, is behavioral isolation. This sometimes has been called "psychological isolation," or "sexual isolation." This type of isolation is a phenomenon which is strictly characteristic of animals and has no counterpart in the plant kingdom. The mating behavior of most animals is complex and involves a series of closely coordinated behavioral activities which reinforce one another and culminate in mating. If, at any stage in this series of coordinated events, the proper response is not elicited by a particular stimulus, then the ritual breaks down and mating does not occur. The choice usually, though not always, is exercised by the female. Males usually exhibit a general lack of discrimination that sometimes leads to ludicrous attempts at mating with other males or with females from very different species. Courting patterns obviously involve a number of types of sexual stimuli based on the senses of sight, smell, hearing, and touch. The colorful plumage in birds, mating calls in birds and amphibians, male attractant odors in moths, smell, or its equivalent antennary sense in *Drosophila* and other insects, wing fluttering in insects, and feather ruffling in birds are all components of sexual recognition and courtship. Experiments have been carried out to determine which senses are most important in the mating behavior of various organisms, and some success has been achieved in analyzing its components. For instance, *Drosophila pseudoobscura* females, who exercise the choice of mating partners, will lose power to discriminate between *D. pseudoobscura* males and males from the sibling species *D. persimilis,* if the females' antennae are removed. If the wings of *Drosophila* males are removed, they are unable to excite the females to copulation. A few species fail to mate in the dark, indicating that visual stimuli are essential. In lightning bugs, the blinking frequency is a critical factor in attracting a male of the same species. In general, these experiments have shown that mating behavior requires a number of different coordinating stimuli all of which are essential to ultimate sperm transference.

Ethological isolation and the other prezygotic mechanisms are much more efficient in isolating individuals, in terms of expended energy, than postzygotic mechanisms, in which offspring actually may be produced. Another peculiar feature of ethological isolation is that while it may work beautifully under natural conditions, it very often breaks down under abnormal ones such as are found in the laboratory. This indicates, for one thing, that ethological isolation is not infallible, at least in closely related species, and for another, that it is not independent of the environment.

Another prezygotic mechanism is mechanical isolation, which appears to be very important in many arthropods, including insects. Mechanical isolation can involve differences in body size, but more often involves the size and morphology of the external genitalia. In insects, the differences in these are so characteristic that they constitute a major taxonomic characteristic.

The last prezygotic mechanism we will mention involves the failure of sperm transfer, the failure of gametes to survive in alien reproductive tracts. One of the best documented cases of this in animals, the failure of sperm survival, is the insemination reaction in *Drosophila* discovered by Patterson (Patterson and Stone, 1952). Normally the sperm are deposited in the vagina of the female from where they are transferred to the various storage organs. In crosses between different species there often is a severe lumping of the semen and fluids secreted by the walls of the vagina. Many times this reaction mass blocks the vagina for several days or more, and signals the failure of the mating. The strength and duration of the insemination reaction varies with the degree of relationship between the species, or strains being crossed. Similar mechanisms probably occur in other animals but have not been well studied. In plants a comparable reaction occurs in which the pollen of certain species (or even certain genotypes, as in the self-sterility alleles in the clovers) fails to grow on the stigmas of others. This kind of phenomenon is sometimes referred to as "physiological isolation."

Postzygotic mechanisms include the failure of zygotes to develop, the lower viability of the hybrid individual (if it survives at all), lower fertility, lower fecundity, or a breakup of coadapted gene complexes by recombination in the hybrid, giving rise to ill-adapted individuals.

Postzygotic mechanisms are more wasteful of the energy resources of the species, and they are most wasteful when hybrid offspring actually are produced which are in varying degrees inviable or infertile, or which themselves produce ill-adapted offspring. There may be several causes of hybrid inviability, but the major one is simply a genetic inbalance giving rise to physiological or developmental disturbances. The inviability of hybrids will obviously affect their fertility. On the other hand, many instances are known in which the hybrids are as viable as the parental species, or even more so, but are partially or fully sterile. The mule is a good example. In such cases chromosome differences leading to meiotic difficulties are usually, if not always, responsible for the sterility. Often there is not sufficient homology between the chromosomes for them to segregate regularly at meiosis.

Under conditions which favor the genetic differentiation of populations, enabling them to become adapted to different environments, there would be considerable advantage in further developing mechanisms which give rise to reproductive isolation between members of different races. Hybrids often will be at a disadvantage in either of two parental environments (hybrid inviability, or weakness). When this is true, selection will favor any differences in physiology, behavior, etc., which prevent mating between the two populations (sexual isolation). Parents which produce these inferior hybrids will be selected against because they will have contributed their genes to the ill-adapted hybrid. This means, essentially, that their genetic contribution to the following generation has been wasted. This surely must be one way in which prezygotic reproductive isolation is strengthened. The plausibility of this assertion has been shown by Koopman (1950) in an experiment in which he manually eliminated hybrids between *D. pseudoobscura* and *D. persimilis* that occurred in a cage population. The number of interspecific matings declined over generations. However, it is also quite probable that in some cases geographical isolation persists long enough for the populations to diverge to the point where there are so many genetic differences between them that reproductive isolation is incidental. The hybrids are genetically unbalanced and don't survive.

Also, the involvement of chromosomal aberrations as isolating mechanisms should not be overlooked. Patterson and Stone (1952) have presented extensive documentation of chromosomal differences between closely related species of *Drosophila*. These differences mainly involve chromosome inversions, as can be seen in Fig. 8.5, but they sometimes involve translocations as well. The translocations, when they occur, are almost always of the centric fusion type which results in the conversion of two acrocentric chromosomes into a metacentric chromosome virtually twice as large (see Fig. 4.1). We have already discussed inversions in relation to supergenes, but the point to be repeated here is that hybrid individuals which are heterozygous for translocations and inversions can be partially sterile because they produce chromosomally unbalanced (aneuploid) gametes (Chapter 4). Chromosomal isolation represents an interesting contrast to the other isolating mechanisms

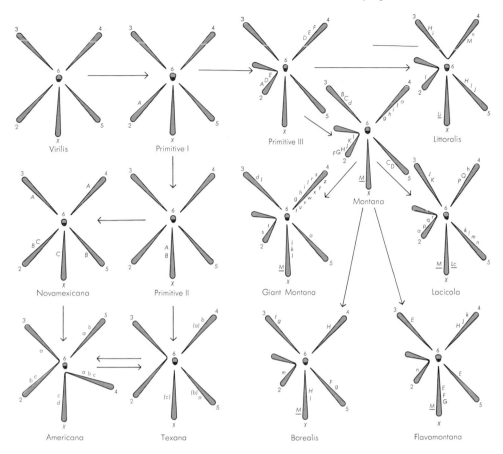

Fig. 8.5. Chromosome evolution in the virilis group [after Hsu, *Univ. Texas Publ., 5204* (1952)]. Capital letters indicate inversion fixed in each species; lower case italicized letters indicate inversions sometimes present. Arrows show sequence used. Each species shows all inversions that were indicated as homozygous earlier in sequence or has exceptions indicated. From W. S. Stone, *Cold Spring Harbor Symposia on Quantitative Biology,* 20 (1955), 260.

we have discussed. A chromosomal aberration is a rare event that arises in a single individual. To survive at all the individual must have some selective advantage at the outset, but beyond that, those possessing the different chromosome will not be very different genetically from their neighbors. This is the reverse of our previous examples, in which genetic differences gave rise to reproductive isolation. It may be that chromosomal differences play a limited role in species formation, however, because of the genetic similarity of the two chromosomal types. Under these conditions they would be in direct competition, with the probable elimination of one or the other.

The widespread occurrence of chromosomal differences in some species, such as *Drosophila,* has sometimes been taken as an indication that they are indispensable factors in speciation. That this is not true has been clearly shown in the Hawaiian *Drosophila* (Carson *et al.*, 1967). On these islands there is a large number of species and the amount of divergence between them is extreme, yet there are virtually no chromosomal differences.

Even though chromosomal isolation is not essential for evolution it may well play an important role in stasipatric speciation. A new chromosomal type arising at one point in a species range could take over large areas until it finally encountered populations which were different enough genetically to be better adapted in their own area, thus stopping its spread. This would give rise to parapatric species.

At any rate, we can see that there are many ways in which reproductive isolation occurs, and the important point is that these give rise to genetically independent lines of descent.

Marginal isolates and speciation

A species range is constantly shifting and moving, its border advancing and receding, amoeba-like, at the limits of the range as conditions permit advances or force retreat. It is thought that such marginal populations may provide especially favorable opportunities for the formation of new species (Carson, 1959; see also Mayr, 1963). Under these conditions, new geographically isolated populations will be constantly produced which have been called "marginal isolates," or "founder populations." (This is shown in Fig. 8.3, item 3, and in this case the geographic barrier will usually be distance.) These will be relatively small and minimally adapted because the species has not been exposed to selection in the new area, not having inhabited it previously. There will be strong selection, though, for adaptation to the environmental niche which the new area provides. This process doesn't differ in principle from the one in our previous discussion of geographic isolation, in which a species breaks up into several large groups in the process of race formation. It differs in detail in that the populations are smaller and thus are more likely to be subject to drift (Chapter 6) and increased homozygosity through inbreeding (Chapter 3), and also in the great frequency with which such isolated populations can occur. Marginal isolates doubtlessly are constantly being formed, while the central bulk of the species remains well adapted to the conditions for which it has been selected. In this way, endless attempts to inhabit new niches can be made without jeopardizing the existence of the species.

This provides also an opportunity for what Mayr has called a "genetic revolution." This involves a very rapid and extensive change in the genetic constitution of the population which can sometimes be extensive enough to give rise to the type of evolutionary novelty that produces new genera and other higher taxonomic categories. Several factors may contribute to the revolution. First, the genetic composition of the founder colony is quite likely to be somewhat different from that of the parent colony. This can be due either to chance or to the

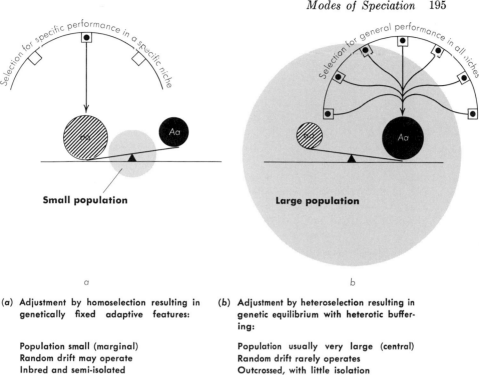

(a)

(b)

(a) Adjustment by homoselection resulting in genetically fixed adaptive features:

Population small (marginal)
Random drift may operate
Inbred and semi-isolated
Open combination usually prevails
Species formation is promoted

(b) Adjustment by heteroselection resulting in genetic equilibrium with heterotic buffering:

Population usually very large (central)
Random drift rarely operates
Outcrossed, with little isolation
Restricted recombination often prevails
Species formation is retarded

Fig. 8.6. Contrasting types of genetic adjustment. Under conditions such as exist in central populations of a geographically differentiated species (b), the scale usually tips in favor of the heterozygote at a particular locus because of its better general performance in all niches (*heterotic buffering*). Because the population is large and outcrossed, it can afford the luxury of continually producing homozygotes which are less fit than the heterozygotes. Small populations (e.g., marginal), because they are too small or too inbred to afford adjustment by heterotic buffering, show the reverse tendency (a). Selection favors the homozygote and leads to the fixation of specific adaptive features. The latter process promotes and the former retards the formation of species. From H. L. Carson, *Cold Spring Harbor Symp. Quant. Biol.*, **24** (1959), 89.

possibility that only particular genotypes in the original population have the capacity to inhabit the new area, even on a minimal basis. On grounds of numbers alone, the smaller isolate will have less genetic variation than the parent population. Whatever variation is present, furthermore, may be additionally reduced by inbreeding. Mayr's major point is that the adaptive value of a gene is influenced not only by direct changes in the external environment, but by changes in its genetic environment as well. This means, then, that the entire genetic system of the isolate will have to become readjusted until it again attains a genetic equilibrium. Furthermore, the tendency toward homozygosity will permit selection for favorable homozygous genotypes that are precisely and narrowly adapted to more restricted, and therefore more specific, environmental conditions. In the parental population, heterozygosity prevails. This results in the buffering of each individual for general performance in all niches. Carson has introduced the terms "homoselection" to describe the situation in marginal isolates, and "heteroselection" to describe selection for the more highly buffered condition which exists in the main body of the population. Most authors suspect that the highly heterozygous, highly buffered parental population is too stable to undergo speciation. Homoselection in small isolated populations thus favors speciation, whereas heteroselection retards the process (see Fig. 8.6). It also has been suggested that genetic revolution is one process that could occur in natural populations and give rise to major changes rapidly. Some paleontologists have been troubled by the absence from the fossil record of common ancestral species which have given rise, in the course of divergences, to higher taxonomic groups, and by the suddenness with which major groups appear. The theory of genetic revolution is consistent with these observations, since major changes are expected to occur rapidly and, in small isolated populations, before they become sufficiently widespread in distribution to be found in the fossil record. Once established, a marginal isolate can begin to expand and to experiment with buffering systems and polymorphisms that are characteristic of well established, highly adapted species. Carson has also suggested, on the basis of studies with several species of *Drosophila* (especially *D. robusta*) that the homoselection in marginal isolates extends to inversion complexes. This breaks down the restrictions on recombination that are operating in the parental populations. Their balanced polymorphisms, based on coadapted gene complexes and maintained by inversion heterozygosity, are not suited to the new area. Homoselection gives rise to populations referred to as being recombinationally "open," whereas heteroselection produces populations which are recombinationally "closed." These terms refer to the relative amount of recombination which can occur. The "openness" of the marginal isolates then permits greater recombination, which increases the efficiency with which new genotypes can be constructed for adaptation to the new environment.

Adaptive surfaces and interdemic selection

So far in our discussion we have limited our attention to selection within populations. This is the aspect of population genetics that has received most of the attention to date. Stabilizing selection, balancing selection, directional selection, normalizing selection, canalizing selection, and disruptive selection are all terms that refer to differential survival of individuals *within* a population. An aspect of evolution that has been largely ignored is the differential survival of populations themselves. The small local breeding population, the deme, is the one which is most important in this regard, and its genetic structure is an important factor in its survival. This would be especially true of marginal isolates. Lewontin (1965) has provided considerable insight into this phenomenon which we now refer to as *interdemic* selection, to distinguish it from intrademic selection. The concept of interdemic selection can to some extent explain certain situations which seem paradoxical when considered only in light of intrapopulation selection. For instance, there are a number of traits which have high individual fitness, and are selected for within a population, which may endanger its continued existence. Body size is one. Generally there is selection within a population for increased body size. However, if there continues to be an increase in size the population could well exhaust its food supply and become extinct. In this respect there can be interdemic selection, with those demes which have not outgrown their food supply surviving. In this way the intrademic tendency toward increased size may not be realized.

For interdemic selection to work well there must be differences between demes with respect to the selective optimum for various traits. Lewontin has shown, with a mathematical model, that this can occur. To take a simple case, consider a polymorphic system of two loci, with two alleles at each locus. There are nine possible (lumping double heterozygotes) genotypes, $AABb$, $AAbb$, etc. Each genotype has a fitness which is denoted by W_i, with i simply indicating that it is some particular genotype under consideration. The frequency of this genotype will be determined by the gene frequencies p_1 and q_1, and p_2 and q_2 at each of the two loci. Its frequency is denoted by Z_i. Thus an average fitness, $\bar{W}$ will be equal to $\Sigma Z_i W_i$, and can be calculated. For the mathematical model one simply assigns individual fitnesses to each genotype of which there are nine in this example. Then the average fitness of a population is determined by the relative frequencies of the various genotypes, which in turn are determined by the gene frequencies at the two loci. If $\bar{W}$ is calculated for a large number of different gene frequencies (this is best done by computer) one gets an "adaptive surface" like the one seen in Fig. 8.7. This is not quite the same as the adaptive field in Fig. 8.1, because now each point represents a population instead of a genotype. Each point on this surface represents a $\bar{W}$ for a particular set of gene frequencies which can be read from the ordinate and abscissa. It should be pointed out that the fitnesses of the different genotypes need not be arbitrarily chosen, but in some populations can be experimentally determined. In Fig. 8.7 this has been done for two inversions (which we can treat as genes in our model) in a grasshopper population, *Moraba scurra*, studied by Lewontin and White (1960).

The most important point with regard to these adaptive surfaces is that they consist of a series of peaks and valleys. Furthermore, they are not all the same height or depth. On this surface a population, which will be represented by a point on the surface, will tend to move "uphill." This represents the action of selection increasing the average adaptive value of the population. In going uphill the population will come to a peak, but it may not be the highest. Consequently a particular deme may find itself on a submaximal peak from which it cannot be moved by intrademic selection. This is because moving "downhill"

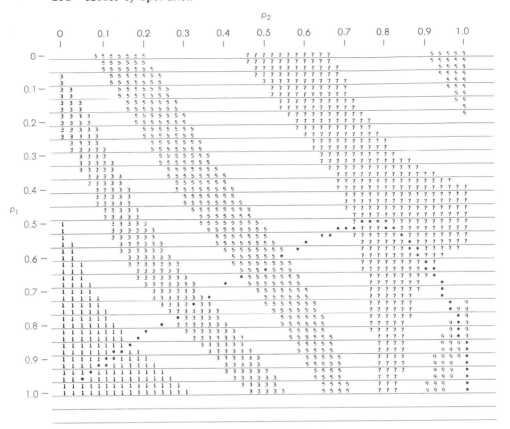

Fig. 8.7. A topographic map of a two locus adaptive surface. In this figure the topographic contours (see Fig. 8.1) have been printed out by computer. The two axes represent the gene frequencies at each of the two loci. The printout numbers represent the population fitness for each combination of gene frequencies. They are computed from the genotype frequencies and the relative fitness values for each genotype. In this case the following values were empirically determined for a real population in which *aaBb* was arbitrarily chosen as 1 before the relative fitnesses were known.

	bb	Bb	BB
aa	0.79	1.0	0.834
Aa	0.67	1.006	0.901
AA	0.657	0.657	1.067

The population with the lowest average fitness consists of all *AAbb* individuals ($p_1 = 1$, $p_2 = 0$) and has a fitness of 0.657. The highest average fitness is that of the population with only *AABB* individuals ($p_1 = 1$, $p_2 = 1$) and is 1.067. The range of fitness values (1.067 − 0.657) has been divided into ten equal portions. Thus, the number 1 displayed on the map represents fitness values from 0.657 to 0.657 + 0.041, the number 2, values from 0.698 to 0.739, etc. The even numbers have been omitted from the map for visual clarity. The adaptive surface pictured here has its lowest spot in the lower left corner. The surface moves uphill diagonally to the right to a height of 7, which is a ridge area since the surface then falls off again toward the upper right corner. The ridge goes uphill to the lower right corner, the highest point on the surface. (Actually, a finer gradation of contours would show that the ridge of sevens is a saddle, rising to slightly higher ground in the upper center also.) This map can also be used to show the changes in gene frequencies in a popula-

from the peak to reach a higher peak involves a decrease in fitness which is opposed by stabilizing selection. With regard to body size, then, or any other trait we choose to consider, some demes could be on submaximal peaks so that a further increase in size is not possible without passing through a selective valley. The stabilization of populations on submaximal peaks for various traits provides interdemic differences on which interdemic selection can act.

Speciation through polyploidy

Polyploidy is the increase in the number of chromosome sets. In higher animals, at least, it is a rare phenomenon. In plants, however, it occurs rather frequently and has been an important factor in plant speciation. (For a more thorough coverage, see Stebbins, 1950). For convenience, a distinction often is made between autopolyploidy and allopolyploidy. Autopolyploids are formed by the multiplication of the number of chromosome sets within a species. Homologous chromosomes occur in sets of four in an autotetraploid, and pair in meiosis as quadrivalents. The multivalency often results in meiotic difficulties, leading to partial sterility, but the mechanisms causing this are beyond our present concern (see Carl P. Swanson, Timothy Mertz, and William J. Young, *Cytogenetics*, in this series). The most common means of polyploid formation are through somatic doubling of the chromosome number in a branch or flower, or through the occasional production and union of unreduced gametes. In self-fertilizing species, somatic doubling in a flower is sufficient. The gametes will be $2N$ and the zygotes $4N$. In cross-fertilizing species, flowers in two different plants can undergo somatic doubling and, if by chance $2N$ gametes from each unites, a $4N$ zygote will be formed. A $2N$ gamete uniting with a haploid gamete will produce a $3N$ gamete. Otherwise, the even more improbable union of two unreduced gametes ($2N$ gametes produced through an aberrant meiotic division) is required. Although tetraploids are common, even higher multiples of the basic chromosome number occur frequently. Autopolyploids usually are quite similar to the diploid species, only larger. One feature of autopolyploids that may have some evolutionary significance is that they are quite frequently reproductively isolated. For instance, a hybrid between a tetraploid and a diploid will be triploid. Triploids are notoriously sterile due to the fact that

---

tion under the influence of selection. This is represented by the path of asterisks. The initial population, $p_1 = 0.95$ and $p_2 = 0.05$, has a low average fitness. Under the influence of selection the gene frequencies change, giving an increased average fitness. The increase from the ones, through the threes, fives, and sevens shows p_1 decreasing and p_2 increasing; but when p_1 begins to increase, the population average fitness increases to the maximum value at $p_1 = 1$ and $p_2 = 1$. (However it might also have gone to the other high point in the upper center.) The change in gene frequencies under selection can thus be seen to lead always uphill. However, the path is not direct and the gene frequencies do not always change by the same amounts or in the same direction, and as was pointed out in the text this could result in the population coming to rest on a submaximal peak. Data from Lewontin and White (1960).

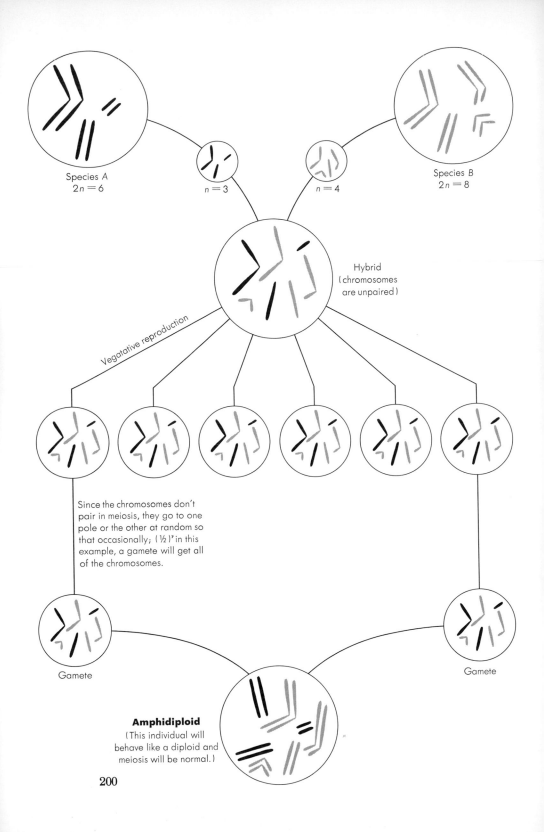

Species A
$2n = 6$

$n = 3$

$n = 4$

Species B
$2n = 8$

Hybrid
(chromosomes
are unpaired)

Vegatative reproduction

Since the chromosomes don't
pair in meiosis, they go to one
pole or the other at random so
that occasionally; $(½)'$ in this
example, a gamete will get all
of the chromosomes.

Gamete

Gamete

Amphidiploid
(This individual will
behave like a diploid and
meiosis will be normal.)

Fig. 8.8. Amphidiploid formation.

three homologous sets of chromosomes give rise in meiosis to anenuploid gametes (gametes with unusual numbers of chromosomes). In such a cell, the homologous chromosomes will occur in sets of three, rather than in pairs as is normally the case. The best segregation that can be achieved by a set of three chromosomes is for one of them to go to one daughter cell and two to the other. Since different sets of three will segregate independently of one another, the secondary gametocyte will have a mixture of single chromosomes and pairs, and only rarely $[2(\frac{1}{2})^n$, where n equals the number of chromosome pairs] will an evenly balanced haploid (n) or diploid gamete $(2N)$ be formed. Thus a tetraploid is another instance in which reproductive isolation could occur before genetic divergence had begun. Furthermore, this represents a means by which the number of gene loci can be increased.

Allopolyploids result from the hybridization of different established species. The hybrids usually are sterile because the chromosomes have not retained sufficient homology to pair well and meiosis is erratic. However, if such hybrids can persist by asexual propagation until, as in the case of autopolyploids, somatic doubling can occur or gametes containing all the chromosomes can be formed, a balanced polyploid condition can arise. Many examples of such allopolyploids exist in nature and some have been produced by man. The latter occur in tobacco, *Nicotiana digluta* (from *N. tabacum* and *N. glutinosa*), cotton, *Gossypium hirsutum* (from *G. arboreum* and *G. thurberi*), and *Raphanobrassica*, formed from a hybrid between two different genera, cabbage (*Raphanus*) and radish (*Brassica*). When the species are related distantly enough for all of their chromosomes to have lost their homology, the polyploid will behave cytologically as if it were a diploid (Fig. 8.8). Such allopolyploids are called "amphiploids," or "amphidiploids." In terms of evolution, the important thing about allopolyploids is that they have many characteristics which differ from both parental species and, since they are reproductively isolated, they constitute a new species which combines many of the characteristics of both parental species. In this manner, an entirely new species can arise in just two sexual generations.

The role of hybridization in evolution

Hybrids are the offspring of distinguishably different parents. "Hybridization" is subject to much variation in usage, but in discussions of evolution the term generally is reserved for describing crosses between different populations, races, or species.

Like polyploidy, hybridization between species is restricted almost entirely to plants. When hybridization between species does occur in animals, it involves species which are still very closely related and between which reproductive isolation has not completely developed.

The reasons for its more common occurrence in plants are not known. However, the greater importance of ethological isolation in animals and the relative rapidity and ease with which it appears to be established, contrasted with its complete absence in plants, has been offered as a partial explanation.

We have already noted the importance of hybridization as a first step in the formation of amphidiploids, but it is important in other ways as well. Basically, the evolutionary importance of hybridization is in bringing together diverse genotypes which can result in major recombinational changes involving not only many gene differences but entire coadapted gene complexes. To accomplish such major changes through mutation and gene replacement would be enormously more difficult and time consuming. On a less drastic scale, interspecies hybrids, even if they are only partially interfertile with the parental species, represent a mechanism whereby gene complexes (indeed whole chromosomes) can pass from one species to another.

Under most conditions, hybrids are less well adapted to any particular environment than either parental species. Consequently, even if they are fertile they will be selected against, will fail to become common, and seldom will be observed. If for some reason, there is a major change in the environment (even if it is only local variation), the hybrids may become common. Furthermore, at the outset the hybrids, being relatively rare, will be more likely to cross with one of the parental species than with another hybrid. This will produce off-spring which, on the average, will have three-fourths of its chromosome complement derived from one parental species and one-fourth from the other. Additional backcrosses to either or both parents will give rise to an alteration of the parental form(s) in the direction of the other species. If the amount of gene flow that occurs this way is small, so that the species remain distinct, the process is called "introgressive hybridization" (Anderson, 1949). This is schematically shown in Fig. 8.3*E*. A striking example of this involves two species of columbines, *Aquilegia pubescens* and *A. formosa,* which grow in the high Sierra Nevada. These differ in a number of characters, some of which are listed in Table 8.2. The flowers of the two species are illustrated in Fig. 8.9. Note that *A. pubescens* is erect whereas *A. formosa* is inverted. The ranges of these species do not overlap extensively, with *A. formosa* generally being found at lower altitudes in more moist habitats, and *pubescens* at the drier and colder higher altitudes. However, in some regions, such as the Saddlebag Lake area, they do overlap and some hybrids are formed. This in itself is an interesting phenomenon because of the way in which pollination is accomplished. In *A. pubescens,* with the erect flower, hawkmoths are the pollen vectors. Just at dusk they begin to visit the flowers and to extract nectar from the long spurs with their equally long tongues. The pollen vector for *formosa* is a hummingbird which is able to hover under the inverted flower and

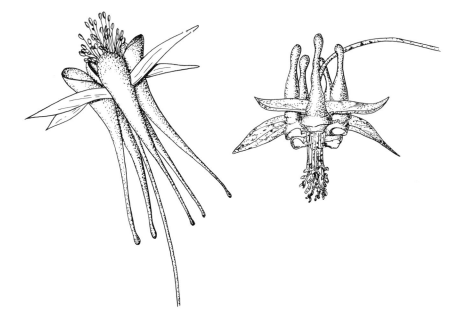

Fig. 8.9. The flowers of *Aquilegia pubescens* (left) and *A. formosa* var. *truncata* (right). From V. Grant, *The Origin of Adaptation*. New York: Columbia University Press, 1963.

Table 8.2 Comparison of *Aquilegia formosa* and *A. pubescens* with respect to five floral characters and their index values

Character	*A. formosa* (*n. Sierra Nevada*)	*Intermediates*	*A. pubescens* (*extreme type*)
Flower position	nodding	horizontal	erect
Spur and sepal color	red	orange, pale orange, pink, or pale pink	pale yellow, or white
Petal blade color	deep yellow	medium yellow	pale yellow, or white
Spur length	10–17 mm	18–28 mm	29–37 mm
Petal blade length	2–4 mm	5–8 mm	9–12 mm
Index value for each character	0	1	2
Total index value	0	5	10

SOURCE: V. Grant, *El Aliso, 2* (1952), 341.

take nectar from the shorter spurs with its beak. Further, the flowers of *formosa* are bright red and yellow, colors known to attract birds, and they are conspicuous during the day. The pale yellow or white flowers of *pubescens* are inconspicuous during the day, to human eyes at least, but become quite distinct in the dim light of the evening when the hawkmoths are about. So hawkmoths seldom visit *formosa*, or hummingbirds *pubescens*. When they do, the moth finds it nearly impossible to get nectar from the inverted flowers, and the hummingbird's beak is too short to get nectar from the deep spurs of the erect flowers. Thus hybrids are infrequent and when they are produced they are seldom visited by either vector. Consequently, the amount of gene flow is small, and there is more of it from *formosa* to *pubescens* than vice versa. In characterizing the adjacent populations in terms of his

Fig. 8.10. Bar graphs showing the frequency distribution of different hybrid index values in three populations of *Aquilegia* in the high Sierra Nevada. These graphs show that introgressive hybridization has occurred between these species, with A. *formosa* genes having passed into the A. *pubescens* populations in this area. From V. Grant, *El Aliso*, 2 (1952), 341.

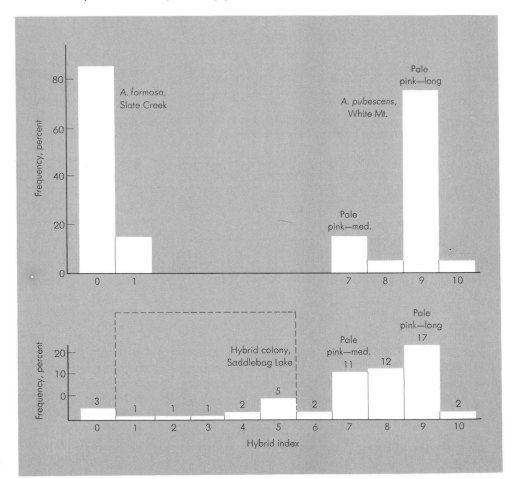

hybrid index (Table **8.2**) Grant found that most of the *formosa* from Slade Creek were of the extreme type, with a hybrid index of zero (Fig. 8.10). In the White Mountain population very few of the individuals were of the extreme *pubescens* type. Most of them had a slight pinkish cast, and scattered throughout the population were medium pink individuals. In regions further removed from the zone of contact the *pubescens* are more predominantly white. Thus we see that limited gene flow can have a definite influence on the characters of the species involved without causing them to lose their distinctive characteristics. The composition of the hybrid population in the Saddlebag Lake area is also shown in Fig. **8.10**.

Sometimes hybridization can spread throughout the entire range of two species whose ranges overlap, leading to a rather complete intergradation between them and giving rise to a single new species. Grant (1963) has called this process in which two species fuse to produce one new one "secondary intergradation" (Fig. 8.3*D*). In this process the original species lose their separate identities.

More recently, Grant (1966) has shown that hybridization is important as a necessary prerequisite for what he calls "chromosomal speciation." In some ways this is similar to amphidiploidy, but it does not involve doubling the chromosome number. The only reported case of this sort occurred experimentally and its occurrence in nature has not been observed. To obtain a chromosomal species, *Gilia malior* and *Gilia modocensis* were crossed. The F_1 were nearly sterile and were intermediate in phenotype between the parental types. The sterility seems to be due to structural differences in chromosomes between the two species which disrupted meiotic pairing. After a number of generations of selection for fertility, it began to increase and then returned to normal. There was a high correlation between fertility and the number of chromosomes that paired. Thus after a number of generations the genome had become stabilized by becoming structurally homozygous for the chromosomes from one species or the other. When the line with restored fertility was tested with either parental type, it was found to be nearly sterile, and thus reproductively isolated. Whether this could occur in nature is not certain because of the low fertility of the first few generations.

In animals, the differences between hybridizing forms are very minor and do not lead to the major evolutionary developments encountered in plants. It can, however, be important in the elimination of genetic differences between previously geographically isolated populations.

Summary

In summary, we have seen in this chapter how gene pools become differentiated and then gradually isolated to become independent lines of descent. This results from the interruption of gene flow between populations to produce races. Races develop into species through further genetic divergence accompanied by the establishment of reproductive isolation.

If anyone doubts the ability of such a system to generate the diversity of life in the world today, let him reflect on the enormity of time during which this process has been at work, while keeping in mind that, given enough time, events that have a low probability of occurrence at any given moment become certainties.

References

Anderson, R., *Interogressive Hybridization*. New York: John Wiley & Sons, 1949.

Blair, W. F., "Mating call in the speciation of anuran amphibians," *American Naturalist, 92* (1958), 27–51.

Blair, A. P., "Variation, Isolation Mechanisms, and Hybridization in Certain Toads," *Genetics 26* (1941), 398–417.

Carson, H. L., F. E. Clayton, and H. D. Stalker, "Karyotypic stability and speciation in Hawaiian *Drosophila*," *Proceedings of the National Academy of Sciences U.S., 57* (1967), 1280–85.

Carson, H. L., "Genetic Conditions Which Promote or Retard the Formation of Species," *Cold Spring Harbor Symposia on Quantitative Biology 24* (1959), 87–105.

Dobzhansky, T., *Genetics and the Origin of Species*. New York: Columbia University Press, 1951.

Grant, V., *The Origin of Adaptations*. New York: Columbia University Press, 1963.

Grant, V., "The origin of a new species of *Gilia* in a hybridization experiment," *Genetics, 54* (1966), 1189–99.

Heed, W. B., and H. W. Kircher, "Unique Sterol in the Ecology and Nutrition of *Drosophila pachea*," *Science 149* (1965), 758–61.

Koopman, K. F., "Natural Selection for Reproductive Isolation Between *Drosophila pseudoobscura* and *Drosophila persimilis*," *Evolution, 4* (1950), 135–48.

Lewontin, R. C., "Selection in and of Populations," in *Ideas in Modern Biology*, J. A. Moore, ed., *Proceedings of the 16th International Congress of Zoology*, Vol. 6. Garden City, N.Y.: Natural History Press, 1965.

Lewontin, R. C., and M. J. D. White, "Interaction between Inversion Polymorphisms of Two Chromosome Pairs in the Grasshopper *Moraba scurra*," *Evolution, 14* (1960), 116–29.

Mayr, E., *Animal Species and Evolution*. Cambridge, Mass.: Harvard University Press, 1963.

Moore, J. A., "Patterns of evolution in the genus *Rana*," in *Genetics, Paleontology and Evolution*, G. L. Jepsen, E. Mayr, and G. G. Simpson, eds. (Princeton, N.J.: Princeton University Press, 1949); available in paperback edition, New York: Atheneum Press, Inc., 1963.

Patterson, J. T., and W. S. Stone, *Evolution in the Genus Drosophila*. New York: Macmillan Company, 1952.

Simpson, G. G., "History of the Fauna of Latin America," *American Scientist, 38* (1950), 361–89.

Stebbins, G. L., *Variation and Evolution in Plants.* New York: Columbia University Press, 1950.

White, M. J. D., "Models of speciation," *Science, 159* (1968), 1065–70.

Wright, S., "The Roles of Mutation, Inbreeding, Crossbreeding, and Selection in Evolution," *Proceedings of the 6th International Congress of Genetics,* Vol. 1 (1932), 356–66.

Wright, S., in *Evolution after Darwin,* S. Tax, ed., Vol. I. Chicago: University of Chicago Press, 1960.

Index